Organizing European Cooperation

Organizing European Cooperation

The Case of Armaments

ULRIKA MÖRTH

ROWMAN & LITTLEFIELD PUBLISHERS, INC.
Lanham • Boulder • New York • Toronto • Oxford

ROWMAN & LITTLEFIELD PUBLISHERS, INC.

Published in the United States of America
by Rowman & Littlefield Publishers, Inc.
A wholly owned subsidiary of The Rowman & Littlefield Publishing Group, Inc.
4501 Forbes Boulevard, Suite 200, Lanham, Maryland 20706
www.rowmanlittlefield.com

P.O. Box 317, Oxford OX2 9RU, UK

Copyright © 2003 by Rowman & Littlefield Publishers, Inc.
First paperback edition 2005

British Library Cataloguing in Publication Information Available

The hardback edition of this book was previously cataloged by the Library of Congress as
follows:

Mörth, Ulrika, 1962-
  Organizing European cooperation : the case of armaments / Ulrika Mörth.
    p. cm.
  Includes bibliographical references and index.
  1. Defense industries—Government policy—European Union countries. 2. Military
weapons—Government policy—European Union countries. 3. Military planning—
European Union countries. 4. European cooperation. I. Title.
  HD9743.E922 M67 2003
  338.47355094—dc21

                                                                          2004299069

  ISBN 0-7425-2807-3 (cloth : alk. paper)
  ISBN 0-7425-2808-1 (pbk. : alk. paper)

Printed in the United States of America

⊖™ The paper used in this publication meets the minimum requirements of American
National Standard for Information Sciences—Permanence of Paper for Printed Library
Materials, ANSI/NISO Z39.48-1992.

To the memory of my father,

Gunnar Mörth

Contents

Abbreviations

ADRIANE	Aerospace and Defense Regional Initiative and Network in Europe
AECMA	European Association of Aerospace Manufacturers
CAG	Competitiveness Advisory Group
CAPS	Conventional Armaments Planning System
CESDP	Common European Security and Defense Policy
CEN	European Committee for Standardization
CEPS	Centre for European Policy Studies
CFSP	Common Foreign and Security Policy
CJTF	Combined Joint Task Force
CNAD	Conference of National Armaments Directors
COE	Council of Europe
COREPER	Comité des représentants permanents
COST	European Cooperation in the Field of Scientific and Tech nological Research
DG I*	Directorate General for External Relations
DG IA*	Directorate General for External Political Relations
DG III*	Directorate General for Industrial Affairs
DG IV*	Directorate General for Competition
DG V*	Directorate General for Employment, Industrial Affairs, and Social Affairs
DG XII*	Directorate General for Science, Research, Development
DTIB	Defense Technology Industrial Base
EAA	European Armaments Agency
EADC	European Aerospace and Defence Company
EADS	European Aeronautic Defence and Space Company

EC	European Communities
ECHR	European Convention on Human Rights
ECSC	European Coal and Steel Community
EDC	European Defence Community
EDEM	European Defense Equipment Market
EDIG	European Defense Industrial Group
EEC	European Economic Community
EMU	Economic and Monetary Union
EPC	European Political Cooperation
ESDI	European Security and Defence Identity
ESPRIT	European Stratgic Program for Research and Development in Information Technology
FT	*Financial Times*
IEPG	Independent European Program Group
EUCLID	European Cooperation for the Longterm in Defence
EUREKA	The European Research Coordinating Agency
IRDAC	Industrial Research and Development Advisory Committee of the European Community
JDW	*Jane's Defence Weekly*
LOI	Letter of Intent
MLG	Multilevel Governance
NAC	North Atlantic Council
NAD	National Armaments Directors
NAMASC	NATO Materiel Acquisition and Support Committee
NAMASP	NATO Materiel Acquisition and Support Program
NATO	North Atlantic Treaty Organization
NIAG	NATO Industrial Advisory Group
NSO	NATO Standardization Organisation
MAS	Military Agency for Standardization
MEP	Member of European Parliament
MLG	Multi-level Governance
MOU	Memorandum of Understanding
OCCAR	Organisme conjoint de coopération en matière d'armement
OSCE	Organization for Security and Cooperation in Europe
PAPS	Armaments Programming System
PARP	Planning and Review Process
PFP	Partnership for Peace
POLARM	Ad Hoc Working Party on a European Armaments

	Policy
RACE	Program for Research and Development in Advanced Communications for Europe
RTD	Research and Technological Development
SAC	Standing Armaments Committe
SCITEC	Study Team Established by Defense Ministers of Nations of the Western Armaments Group with the Participation of the European Defense Industry Group
SDI	Strategic Defense Initiative
SEA	Single European Act
SIPRI	Stockholm International Peace Research Institute
SME	Small and Medium Sized Enterprises
STANAGS	Standardization Agreements
TEU	Treaty on European Union
TDC	Transnational Defence Company
UN	United Nations
UNICE	Union of Industrial and Employers' Confederations of Europe
WEAG	Western European Armaments Group
WEAO	Western European Armaments Organisation
WEU	Western European Union

* In fall 1999, the Commission reorganized and the general directorates changed names and to some extent tasks. In this book, however, the old names of the DGs are used because the empirical analysis is mainly focused on the period before fall 1999.

Preface

This book project started in 1996 when I got a postdoctorate year at Score (Stockholm Center for Organizational Research). A generous funding from what was then the Swedish Agency for Civil Emergency Planning (ÖCB) made it possible for me to stay in Brussels for several months at CEPS (Centre for European Policy Studies). I am grateful to the then-director of CEPS, Peter Ludlow, for his kindness in allowing me to stay at CEPS for various periods in 1996 through 1998.

The project was initially focused on the European Commission and its handling of issues that fall between the first and the second pillars. During my various stays in Brussels I soon realized that one issue in particular raised many fundamental empirical and theoretical questions, namely, the issue of armaments. In 1996 it was still early for the issue in EU's policy-making process. I was therefore able to follow the process by which the issue of armaments was framed and handled by the European Commission. At the same time, as the issue of armaments became more politicized, my research project became larger.

In my analysis of this highly complex and multifaceted issue I had great help from the TREO (Transnational Regulation and the Transformation of the State) Research Group at SCORE, financed by Swedish Council for Research in the Humanities and Social Sciences (HSFR). In the TREO project I was able to combine my knowledge and insights from political science, especially theories on European integration, with theories on institutions and organizations. I am especially grateful to Bengt Jacobsson, Kerstin Sahlin-Andersson, and Kerstin Jacobsson. It was Bengt who convinced me that the organizational field approach was the most interesting way to tell the story of how an issue is the object of parallel and interlinked processes. I am also grateful to Burkard Schmitt, Gunilla Herolf, and Kjell Goldmann for useful comments on early drafts. The ECPR Workshop in Grenoble in 2001 was also helpful to me in finishing the book.

I also wish to thank all the people I have interviewed. Without them, this book would not have been possible. Thank you all; I am deeply grateful. My former supervisor, Bengt Sundelius, has always taken a great interest in my research. He has been very generous with comments and advice on several drafts of the book.

To my colleague and husband, Jan Hallenberg: I do not know how I can thank you enough for your support in writing this book. Since we are in the same line of work I can only say—I will be there for you when you write your next book!

—Ulrika Mörth
Stockholm

Chapter 1
Introduction

Aim of the Study

In the wake of the end of the Cold War, the Kosovo War, the Amsterdam Treaty, and the so-called Cologne process[1] in the European Union (EU), the question of Europe's defense capacity and policy is high on the European political agenda. In the Amsterdam Treaty in 1997 the European defense dimension was strengthened. It was stated, "The common foreign and security policy shall include all questions relating to the security of the Union, including the progressive framing of a common defence policy . . . which might lead to a common defence, should the European Council so decide" (Article 17). The conflict in Kosovo during 1998 and 1999, combined with a British and French military initiative in 1998, started a discussion of the need for a common European defense policy and military capacity. In June 1999, the Cologne Declaration gave a detailed blueprint for strengthening the Union's ability to launch military action. Six months later, the European Council agreed to create a new security and defense capability, which would allow the EU to take military action. In the Nice Treaty this political ambition was reaffirmed, and it was stated that "the objective for the European Union is to become operational quickly" (TEU Article 25). This was followed, in the presidency conclusions of the Belgian chairmanship at Laeken, by a declaration on the operational capability of the Common European Security and Defense Policy (CESDP).

One important issue in the discussion of the formation of a European defense capacity and policy is armaments. Indeed, without a strong European defense industry and cooperation on armaments, a European defense capacity and policy will have no substance. Recent years of defense industrial restructuring and consolidation at the European level, combined with various intergovernmental initiatives, have shown that the issue of armaments is the subject of a vigorous European cooperation.

For those familiar with the history of the European integration process, the recent development on defense issues within the EU is reminiscent of the political efforts in the 1950s to create a European Defense Community (EDC). While the political vision of a European defense capacity was very strong during the early days of European integration, the vision did not become a reality then. Defense issues were handled outside the European Union after the political failure of the EDC.[2] Hence, the issue of armaments played a rather shadowy role in the EU until the political breakthrough for defense issues in the late 1990s and early part of the new millennium. However, the issue of armaments is not rooted only in the European defense context; it has also been part of the market-making process in the EU, and part of the classic theme of how Europe technologically and economically lags behind the United States.

Indeed, enduring clusters of issues seem to circulate throughout the European integration process. They travel both in space and in time (cf. Merton 1965). Issues may change in their specific forms, but the core tends to remain the same (Merton 1965). We should therefore focus on processes of attention rather than on new issues. Indeed, various EU policy initiatives and proposals can be traced to earlier periods in the European integration process.[3] Since issues travel not only in time but also in space, the same issue can be the subject of attention in different policy processes. An issue can at one point be discussed in one specific policy process and, later, in another policy process. It is also possible for an issue to be discussed simultaneously in various policy processes.

The issue of armaments thus belongs to different European projects—the political economy project, developed through the EU, and the defense and security project, organized through NATO and the Western European Union (WEU) and recently through the EU. This means that the issue of armaments has been conceptualized in different ways. Indeed, issues do not travel as ready-made packages. Different conceptualizations and categorizations of the same issue can compete with each other. Competition between conceptualizations and definitions is a crucial feature of all politics. This is particularly true for the European Union, in which making policy is an exercise in the mobilization of ideas and policy conceptualizations (Peters 1994; Mörth 2000a). Interesting questions are why particular conceptualizations appear and why one becomes authorized rather than another. Furthermore, why is it that an issue is considered to be politically important at a certain point in the European integration process?

Various actors within the first (the community pillar) and second pillars (Common Foreign and Security policy [CFSP[4]]) of the EU have claimed the right of definition and categorization of the issue of armaments. Parts of the Commission have argued that the issue of armaments belongs to the community pillar, whereas the EU governments have claimed that the issue is a building block in the EU's emerging defense policy within its second pillar. The controversy and tension between these two ways of categorizing the issue of armaments concern fundamental questions in European politics. To which pillar of the European Union does the armaments issue belong—to the first pillar and the supranational character of EU decision making, or to the second pillar (CFSP) and an intergovernmental policy-making style? How many of the legal rules of the internal market can be applied to the issue of armaments, and to what extent can national governments protect their defense industries from the market-making activities?

The traditional boundaries in European politics have, in recent years, been questioned regarding the issue of armaments since it is dependent on both projects of European integration. Indeed, European cooperation on armaments is perceived to require cooperation between the market-oriented organization of the EU (the community pillar) and the defense-oriented organizations of the EU (the second pillar), the WEU and its subsidiary organs the Western European Armaments Group (WEAG)/Western European Armaments Organization (WEAO), as well as NATO. This opinion has most clearly been put forward by the European Commission and the European Parliament, and by EU governments.

The main theme of this book is an analysis of how the issue of armaments has been handled and conceptualized within the two different spheres of integration and how these projects have moved closer with respect to the issue of armaments. It is an analysis of cross-pillar interaction between two organizational fields and the emergence of a new organizational field.

I have focused more on how the organizational fields have moved closer and less on why this has happened. However, an analysis of how

European cooperation is formed necessarily involves studying the dynamics behind this process.

I argue not only that the case in point illustrates the complexity of cross-pillar issues, but also that the borders between the market-oriented sphere and security-oriented sphere in general are diffuse and difficult to uphold. The empirical question concerns how this development of blurred borders between the two projects should be interpreted. Are we dealing with overlaps between the market-oriented and the security—oriented projects that force actors to modify their positions, conceptualizations, and legal rules in order to establish a closer European cooperation on armaments? Or does the formation of European cooperation on armaments illustrate more fundamental changes in the conceptualization and handling of issues that have market and security dimensions? To put it briefly: Does the case of armaments entail European cooperation based upon lowest common denominators, or is it based upon a common understanding of how to deal with the issues that have market and security considerations?

European Integration as Organizing

A traditional analytical starting point in a study on the formation of a European defense- and security-related policy would be an analysis of national interests and the intergovernmental pooling of sovereignty in the European Union. This state-centered intergovernmental approach can be regarded as a standard operating procedure in the foreign, security, and defense-related literature. This book presents a different analysis of European politics, taking European organizing as its starting point. This does not mean that states are unimportant in the analysis. They can be studied as organizations as well. The perspective presented in this book is not, however, that of separate states having positioned themselves on the issue of armaments, and the focus is not on the ways in which these states have pooled sovereignty in an intergovernmental bargaining process. Instead, states are part in the analysis in a collective way when they act together in the EU Council of Ministers or in another organizational form at the European level.

In some respects I conduct an intergovernmental study, but my view of how states are part of the European integration process differs from the intergovernmental theoretical perspective on the EU and European integration (Moravcsik 1998). The intergovernmental approach is based upon a traditional Weberian state concept and is therefore focused on the organization of authority within the state (Krasner 1999). It is also assumed that governments control the political process and that they are masters of the treaties. According to an intergovernmental analysis it is therefore logical to study how states—especially the governments and the top politicians—negotiate at the European level. However, I would argue that you can also analyze states from a process perspective in which you see them as part of the formation of European authority structures. Furthermore, they are not alone in this process. Governments may lose their grip on power in an integration process, as is argued by the students of multilevel governance (Marks et al. 1996). They are increasingly interacting with various transnational actors such as European industry, the European Commission, the European Parliament, and the European Court of Justice (Sandholtz and Stone Sweet 1998). Governments have to interact with each other in different intergovernmental bodies and they have to make joint decisions. The intergovernmental organizations are not merely an aggregation of national preferences. The Council of Ministers and other intergovernmental organizations are actors in their own rights and should be studied not only from a national perspective but also as part of the ongoing organizing taking place at the European level.

Drawing on Ruggie's notion of the EU as a "multiperspectival polity" it can be argued that states cannot be treated as external actors in the EU (Ruggie 1993). This is so because complex transformative processes challenge the Westphalian state system. States are "endogenous to the preferences and positions of all involved, making it impossible to distinguish 'us' and 'them' for all practical purposes" (Smith 1996, 13). The embeddedness of the state—the government and the administration—is crucial to my understanding of European politics (Mörth 1996, 1998). This means that national administrations are part of the WEU, NATO, and the EU. To what extent states are embedded in these organizations is an empirical question. I argue, however, that concepts and theories guide empirical analysis and that we often find what we seek. Traditional intergovernmentalism is focused on how states use the EU as an arena for intergovernmental bargaining. It is therefore most likely that the approach limits the empirical analysis in the sense that it is difficult to study a more endogenous state preference formation process. Another state concept is needed to study that process.

In contrast to the Westphalian state, states can be regarded as disjointed and fragmented. "Process and activity become more important than structure and fixed institutions. The state becomes not so much a thing ...as a set

of spatially detached activities, diffused across the Member States" (Caporaso 1996, 45). According to this reasoning, states can still be regarded as
important actors in the integration process. The problem with the traditional intergovernmental analysis is that it does not cover new types of state
activities and new forms of authority structures. A better term to capture
a more intense and extensive state engagement is "intense transgovernmentalism" (Wallace and Wallace 2000). Intense transgovernmentalism
entails governments being committed "to rather extensive engagement"
(Wallace and Wallace 2000, 33), but this commitment is not necessarily
based on hierarchical supranational decisionmaking and coercive rules. It
can also consist of a more governance-like authority structure that is based on
soft law, networks, competition, and knowledge (cf. Boli and Thomas 1999).

I include states in the analysis in the way that I see states—governments and administrations—as participating within the organizing process
toward a European authority structure. States are thus inside the organizations in the organizational fields and are not treated as external actors.

One crucial motivation for using the concept of an organizational
field in a study of European integration is that it enables us to include in
the analysis all organizations and organizing activities that operate in the
same domain and critically influence each other's performance. It is not
necessary to categorize the issue at stake a priori; rather, the analysis can
be based on the organizational activity that takes place on an issue irrespective of whether the organizations are the European Commission, the
Council of Ministers, or an interest organization. Furthermore, I analyze
European cooperation as a process of organizing and not as a process of
integration. The main reason for this is that the concept of organizing is
broader than the concept of integration. Integration has traditionally been
defined in terms of a process that leads to a new form of political community and center (Haas 1958/68; Lindberg 1963). This means that students
of the EU and European integration are focused on the establishment of
formal organizations and on a government structure. I would, however,
argue that the lack of formal organizations, hierarchical relations, and formal transfer of sovereignty within a policy area cannot automatically be
interpreted as a weak process of institutionalization (Brunsson 1999). The
concept of organizing, and the literature on how organizations are institutionalized, covers both a government structure (formal organizations) and
a governance structure (informal organizations). I sometimes use the term
European cooperation to cover various forms of organization and organizing activities. However, in contrast to the concepts of organizations and

Introduction

7

organizing, European cooperation lacks theoretical and analytical potential.

Another, related, theoretical motive for using an analytical model from the literature on organizations is that sociological institutionalism emphasizes rule-based behavior. This approach makes it possible to highlight different and somewhat new components and stories of the European integration process, which is not the case with the traditional theories on European integration. I have already raised some critical points about the intergovernmental perspective on the EU and European integration. Let me develop this critique further and also explain why the other salient theoretical perspective on the EU and European integration—neofunctionalism—is not applied in the book. In my work on the empirical process I have come to the conclusion that although the European integration literature is a natural starting point in the analysis, it also overlooks fundamental aspects of European politics. Paradoxically, the literature on the EU and European integration seldom focuses on the organizations and processes of organizing, although organizations and organizing lie at the very heart of the European political cooperation process.[5] The main theoretical perspectives on European integration, intergovernmentalism, and neofunctionalism focus on various strategic actors (national governments or supranational actors), but these actors are seldom analyzed as part of a wider organizational and institutional context. Actors and policies seem to float freely. The actors are driven by their strategic interests, and neither theoretical perspective analyzes how these interests are formed. Although some of the organizational complexity can be recognized in the two analytical perspectives (that is, multiple actors strive for political influence), the interaction between the actors and the institutional setting—the rules of the game—is seldom problematized. The rational actor model is crucial in both theoretical perspectives, which give no space for an analysis of how values, norms, and rules are created. Furthermore, students of European integration often assume that the power struggle between the intergovernmental and the supranational paths of the integration process is a zero–sum game.

My view of the EU and the European integration process is therefore similar to that of multilevel governance, which, among other things, questions the notion of a zero–sum game between the intergovernmental and supranational political levels. The notion of the EU as a multilevel governance (MLG) organization takes into account the complexity between different actors at different levels, but it seems to have left out the important discussion of how some relationships are more important than others. Even though the MLG perspective assumes that authority and policy-making influence are shared across multiple levels of government (that is, subnational, national, and supranational), the analysis of power is not always convincing (Marks et al. 1996). An organizational field perspective on European integration, on the other hand, analyzes power relations between organizations and how formal/informal domination and authority structures are formed. The conclusion that I draw from this is that MLG is not yet a fully developed theoretical perspective on the EU and European integration. The theoretical potential could, however, be rather large. In the final chapter I return to this discussion of how organizational theory and theories on European integration can enrich each other.

An empirical problem with the MLG perspective is that students of MLG focus only on first-pillar questions. The lack of cross-pillar considerations leads me to a fundamental problem in the literature on the EU and European integration, namely, that complex relations between different paths of European integration are rarely analyzed. Indeed, there are few studies that focus on the linkage between the community pillar, for instance, the creation of the Economic and Monetary Union (EMU), and the formation of a common foreign and security policy (Sperling and Kirchner 1997). This is rather remarkable considering that the very beginning of the European integration process, with the creation of the European Coal and Steel Community (ECSC), is considered to have been driven by security motives (Haas 1958/68). "Thus the aims of the ECSC were by no means simply economic; it was intended to address the major security concerns of the early Cold War period" (Bretherton and Vogler 1999, 210).

There are thus two separate literatures and analyses on the EU and the European integration process. First, there is a huge literature on the community pillar, on market deregulation, EMU, research and development, social policy, and so on, that has developed its specific discourse, analysis, and concepts. One dominant theoretical approach is the neofunctionalist analysis of the European integration process, with its emphasis on spillover, the integration as a process, and the focus on supranational and transnational actors (Haas 1958/68; Lindberg and Scheingold 1970; Sandholtz and Zysman 1989; Sandholtz and Stone Sweet 1998). Second, there is vast literature on the so–called security triangle (the EU, the WEU, and NATO) and the formation of a common foreign and security policy within the European Union. This type of analysis of European politics has also developed its own discourse and theoretical approach. The analysis is often based on theories from the literature on international relations, with its emphasis on states and national interests (Eliassen 1998; Peterson and Sjursen 1998; Rhodes 1998; Bretherton and Vogler 1999). It should also be noted that the EU's civilian integration process is studied from an intergovernmental approach (Moravcsik 1991, 1993, 1998),[6] but there is, to my knowledge, no example of a neofunctionalist approach to the EU's CFSP.[7]

One important consequence of the existence of these separate analytical and empirical traditions is that political issues and policy areas tend to be categorized and analyzed within either of these analytical approaches, even though the empirical political process indicates something else. It is likely that most do not even notice that issues seem to move across organizations and different political spheres. How we conceptualize and define the issue tends to vary depending on which academic domicile we belong to and how the issue is presented by media, politicians, and other researchers. By using concepts and analysis from the organizational literature and focusing on organizational activity at the European level, I reduce the risk of telling only one part of the story of the armaments issue. The concepts of organizations and organizational fields will also help me to study empirical complexity without loss of analytical clarity.

The establishment of European cooperation on armaments is analyzed as an organizing process in which institutionalized organizations interact with each other. European cooperation is thus analyzed as a process that gradually takes shape in terms of the organizational setup but also in terms of institutions, that is, rules that guide collective behavior. The concept that is used to analyze the organizing process is that of organizational fields. Organizational fields consist of organizations, which are held together by institutionalized rules. These rules determine how issues are interpreted and categorized. Indeed, how issues are framed and put into a certain political context is determined by institutionalized rules of the organizational fields. This means that issues can belong to different organizational fields in which there are different rules that specify and define the limits of the political agenda and the policy process.

The issue of armaments is part of two organizational fields. The first field is called the defense field. The actors in this field are preoccupied with the EU's security and defense policy. The second field is the market field. The actors in this field focus on the community pillar such as research and technological development (RTD) and industrial matters.

The empirical questions are the following: What are the institutionalized rules in the defense field and how is the issue of armaments framed? What are the institutionalized rules in the market field and how is the issue of armaments framed? The recent development of European cooperation on armaments suggests that the boundaries between the two organizational fields are blurred and that new types of cooperative arrangements on the issue of armaments have emerged. Organizations are not given entities. They are involved in a constant process of organizing, internally as well as externally. Their identities and functions change. This is especially true of the European security architecture after the end of the Cold War. The general picture of the relationship between the EU, the WEU, and NATO is that of unclear competencies and authority relations. This lack of clarity creates frictions among the organizations and goes back to fundamental issues in European politics, for instance the relationship between Europe and the United States. The end of the Cold War has created a turbulent situation in which rivalry and dependence among the organizations are more evident than ever. Thus, the organizations take part in a conciliation process (Aggarwal 1998). Bringing market and defense issues closer to each other is not an easy process; it challenges the traditional separation between "low" and "high" politics and how policy issues within these spheres are conceptualized, regulated, and politically handled. This process is also about identities in the sense that organizations have to develop new roles and self-images in the post-Cold War era (cf. Cederman 2001). They are seeking new identities in a situation where it is difficult to uphold the borders between the market and defense domains. This is problematic for them since identity is formed in relation to others (Stråth 2000).

The changes in the two fields—defense and market—are studied as the formation of a new organizational field in which both fields are included. The overall question in that analysis concerns the coexistence of deadlocks and development in European politics (Héritier 1999). From an organizational perspective, this puzzle can be stated in the following way: How do organizations that compete with each other, and that are also interdependent, cooperate? How do they interact and communicate with each other and in what ways are the two fields moving closer? The starting assumption in the study is that the two fields are moving closer to each other. The empirical question is how thin or thick the emerging organizational field is in terms of institutionalized rules, frames, identities, organizational authority, and power relations among different organizations from the two fields.

Method and Sources

I focus primarily on the European Union and the WEU, WEAG, and WEAO, but I am also interested in other types of European organizing activities related to armaments. This is a book on European cooperation on armaments, and I have therefore included NATO in the analysis only when NATO's activities related to armaments are directly linked to the other European activities related to armaments; that is, if NATO's activities are mentioned in EU/WEU documents or by EU/WEU officials, or if NATO documents contain explicit links to European activities. The material consists of various types of official documents from the EU, WEAG/WEAO, NATO, and other European organizations.

I have also conducted interviews with officials within the EU, NATO, and WEAG/WEAO and with French and Swedish officials, especially from the ministries of defense.[8] The interviews have, in general, functioned as important sources of information on how various actors have perceived the issue of armaments and the questions related to this issue (see an interview list in the appendix). The interviews, in combination with the written material, have given me insights into how the issue of armaments has been framed within the organizations. The interviews have been important sources for my study of how the organizations interact and how they perceive each other (chapter 5). These questions cannot be studied using the written material alone, although documents from both the Council of Ministers within the EU and the loose cooperation within the letter of intent (LOI) process are very detailed with respect to how the governments view European cooperation on armaments and what kind of European organizational setup they would like to see.[9] Furthermore, my interviews have been conducted with officials and not with politicians, except in the case of the Swedish defense minister and three Swedish members of the European Parliament. One important reason for this is that politicians have not been willing to discuss this sensitive issue, which concerns national issues of core sovereignty, at least not during the formative period on which I have been focusing. Another reason is that rather early in my empirical work I realized that the few governmental declarations and decisions on the issue of armaments that exist only codify and legitimize a process that was already under way. Thus, the dynamics of the process seem to have taken place somewhere else. My research has therefore focused on the relationship between the day-to-day integration process and the history-making level (Peterson 1995; Peterson and Bomberg 1998).

My exclusion of separate national actors and perspectives, other than in exceptional cases, will of course mean that the picture of how some of the intergovernmental organizations perceive the European organizing process could be considered incomplete. In the case of the cooperation within the organisme conjoint de coopération en matière d'armament (OCCAR), for instance, there are very few documents from the organization that say anything about how it perceives WEAG, for example.[10] It has also been a problem that WEAG and WEAO are rather loose organizations and that they seem to exist only within national ministries and agencies. Virtually all the individuals from WEAG who were interviewed were working at the secretariat in Brussels rather than with national officials. This means that only parts of the organizations are studied. It is my belief, however, that the people in the secretariats in Brussels convey important images of other organizational setups and of how they interact with other organizations. Indeed, those people within the organizations who were interviewed act on behalf of those organizations (Arhne 1994). The question is, however, whether the people interviewed can be regarded as representatives of their organizations, and if some other persons might have been interviewed who would have given another version of the story in question that would be important for the analysis. This risk is reduced since I rely not only on interviews but also on various types of official, and sometimes unofficial, documents.

The various texts and interviews have all been important in this rather complex empirical process concerning the creation of European cooperation on armaments. By putting many bits and pieces together from my interviews and from my reading of various documents, I have tried to identify the important actors and components in the organizing process, the framing activities, and how various organizations interact with and monitor each other. I have followed this organizing process since 1996, and I have interviewed the same people on several occasions from 1996 through 2000 (see the appendix). The method is a type of process tracing (George and McKeown 1985), although I am not studying specific decisions but, rather, the various steps in the formation of a European organizational field on armaments. I have thus traced how an entire field evolves rather than a single decision.

In my study of armaments I have come across various policy centers such as the Centre for European Policy Studies (CEPS) in Brussels and the WEU's Institute for Security Studies in Paris. These centers have published several reports written by researchers. I sometimes use these reports as sources in the empirical chapters. I also consider the centers to be actors and part of the political process related to the creation of a closer European cooperation on armaments (chapter 5). I have, however, tried to get as many sources as possible and to avoid being dependent on one source only. A similar problem regarding sources concerns the specialized journals on defense matters. These not only serve as reports of industry developments but also contribute to an image of a capability gap between Europe and the United States. In the empirical analysis, special journals on defense issues play an important role as sources, especially in chapter 4. This is because several of the developments analyzed here are difficult to cover in any other way. I try to alleviate this problem in two ways: first I cover several journals; second, I have also conducted several interviews with industrialists and academic specialists as well as officials.

Design of the Book

Chapter 2 outlines the theoretical concepts and analysis of the book: organizations, organizational fields, and frames. Chapter 3 discusses the EU's defense capacity and the framing activities on the issue of armaments within the defense field. Chapter 4 presents the market frame on the issue of armaments and how parts of the European Union (especially the European Commission) are part of an organizational field, which is occupied with the broader discussion of Europe's economic and technological competitiveness toward the United States. The emerging organizational field in which the market–and defense–based organizations interact is analyzed in chapters 5 and 6.

The concluding chapter discusses how an organizational approach can enrich approaches to European integration in our studies of European politics. How have the theoretical concepts in this study helped us to understand the development of European armaments cooperation in ways that European integration theoretical approaches would have missed? Furthermore, what type of analysis is feasible if we want to study empirical processes that belong to different analytical traditions? How can we cross the almost unbreakable border between European market-making activities and the security-oriented political process? Is this distinction a function of an academic division of labor or is it based on explicit (or implicit) philosophical arguments?

Notes

1. By the Cologne Process I mean the undertaking by the EU to establish a force for crisis management, which is to be completed by 2003.

2. The term European Union (EU) is used throughout the book except when I explicitly refer to the pre—Maastricht era.

3. Héritier argues, for instance, that the Commission has an institutional memory. "The Commission has a long institutional memory and a remarkable amount of patience in pursuing policy innovation plans over time. These are accompanied by multiple activities, such as the issuing of Green Papers, and extensive consultations in order to build supportive networks, until the time is ripe for placing the issue on the agenda" (Héritier 1997, 179; Héritier 1999).

4. Due to the ongoing development within the second pillar, new acronyms are used in various documents and by the interviewees. These are CFSP, CEPSD, and EDP. CFSP—common foreign and security policy—was formally part of the Maastricht Treaty. CEPSD—common european policy on security and defense—has been used in the documents of the European Council in Helsinki in December 1999. EDP—European Defense Policy—is sometimes used in various texts as a shorthand for CEPSD (Heisbourg 2000). In this book, I use CFSP since this is the only treaty-based acronym. I have, however, used the other two acronyms when they appear in direct quotations or when they were otherwise used by interviewees.

5. There is some confusion over the concepts of organizations and institutions in the literature on EU and the European integration process. The word *institutions* is used mainly to describe empirically the various EU bodies or organizations. The EU is also analyzed from a new institutionalist perspective (Bulmer 1994, 1998).

6. For an excellent overview of theories of European integration see Rosamond 2000.

7. However, in his book *Foreign Policy in the European Union* (London and New York: Longman, 1999), Ben Soetendorp touches upon some aspects of neofunctionalism in his analysis of EU foreign policy.

8. There are three categories of interviewees. The first includes those who do not wish to be either quoted or referred to. Interviewees in the second category do not wish to be directly quoted but do not object to being referred to. This is the largest category. Finally, there are some who accept even direct quotes. All interviewees are listed in the appendix.

9. Letter of Intent was initiated in 1998 by six European governments with the aim of strengthening cooperation on armaments (chapter 5).

10. In November 1996 the Joint Armaments Cooperation Organization,

OCCAR (Organisme conjoint de coopération en matière d'armement), was created to act as a joint program office on behalf of France, Germany, the United Kingdom, and Italy (chapter 5).

Chapter 2
Theoretical Approach on European Organizing

Organizations

In the organizational literature, two types of organizations are often contrasted with each other. The first type is the formal organization, which is highly coordinated and stable and is established to accomplish certain functions. Such organizations are intentional and rational, and they adapt to the external environment (Brunsson 1999). They are also backed by hierarchical authority, that is, legislation (Brunsson 1999). However, it is argued in the organizational literature that organizations seldom demonstrate these characteristics. Instead, they are often characterized by conflicting goals, heterogeneity, ambiguity, and unpredictability (March 1981; Christensen and Laegreid 1998; Brunsson 1999). Thus, if we open the black box we discover conflicts and loosely coupled units with a polycentric power structure (Brunsson and Olsen 1998). It can also be argued that organizations have a complex relationship with their external environment and that organizational boundaries are vague and permeable (Brunsson 1999). Indeed, organizations seem structurally to reflect socially constructed reality rather than to adjust to changes in the external environment (Berger and Luckmann 1967; Meyer and Rowan 1977/91; Brunsson and Sahlin-Andersson 1997).

Drawing from this critique of a formal organizational conceptualization, a more informal organizational concept has been developed that presents the view that something is organized when there are some forms of coordination and order between actors, individuals, or collectives (Brunsson 1999; Ahrne and Hedström 1999). Organizations are defined as "groups of individuals bound by some common purpose to achieve objectives" (North 1990: 5; Ahrne and Hedström 1999). An organization thus consists of some form of structured interaction and relationship between actors. This means that organizations are conceptualized in terms of processes—organizing—rather than in terms of static entities. It also means that organizing is not dependent on the formal structure. As Brunsson puts it, "formal organizations do not always work in as organized a way as they say they do or as others describe them as doing, while markets are often much more organized than common beliefs about markets would suggest" (Brunsson 1999, 120). Not all organizations are formal organizations backed by a hierarchical and legislative form of authorization and legitimization; some are looser structures based on networks and voluntary agreements. Thus, according to the informal way of conceptualizing an organization, there are many ways of organizing other than those that entail traditional forms of authority and governance. The key argument in this line of thinking is that there exist organizational activities that lie outside traditional formal organizations, and that these informal organizations create rules that are followed. This is the case with organizing through standardization and other types of soft law, which is a form of rule and organizational activity that is not backed by hierarchical authority, that is, legislation. Standards are explicit and voluntary rules, but they are difficult to disregard (Brunsson and Jacobsson 2000).

In practice, the separation between formal and informal organization is problematic if we expect distinct, empirically observable differences between them. This is always a problem with analytical categorizations. Making such a distinction can, however, be a useful analytical categorization to show that organizations can take many forms (cf. Brunsson 1999). This is especially important when analyzing and studying the European integration process, which is characterized by multiple forms of organizational activity (Héritier 1999; Wallace and Wallace 2000). A broad organizational concept that defines organizations in terms of structured interaction and relationships between actors will therefore capture both formal and informal organizations. Indeed, some organizations in the European integration process are treaty-based and have a supranational and rather clear authority structure. An obvious example of a European organization is the European Union, which, in turn, consists of multiple organizations: the European Commission, the European Parliament, the European Council, the Council of Ministers, etc. These supranational and intergovernmental organizations are formal organizations, even though they also exhibit informal characteristics. There are also less formal organizations based on soft law agreements that demonstrate a loose organizational structure. Empirical examples are the European Research Cooperation Agency (EUREKA) (chapter 4), the Schengen agreement (chapter 5) or OCCAR (chapter 5), and the Letter of Intent (chapter 5). The more informal way of organizing can be characterized as governance without a political center and hierarchy (Caporaso 1996; Christiansen 1997, 65, see also Rhodes 1997; Kohler-Koch 1997).[1] A broad conceptualization of organizations is also important to study processes of organizing and organizational dynamics. The formal conceptualization defines organizations as closed entities and as conditions, whereas the informal conceptualization does not draw any clear boundaries between the organization and its environment. The latter also recognizes that organizations are involved in constant processes of organizing, internally as well as externally. These processes of organizing and change do not take place in an ideational vacuum. Organizations do not float freely but take part in various organizational networks.

Organizational Fields and Frames

A key argument in the institutional analysis of organizations is that they are influenced by "widely held norms and ideas about the kind of organisational forms that are natural, correct or desirable" (Brunsson 1998, 260). Organizations also influence the norms, ideas, and rules of the external environment (Ahrne and Hedström 1999; Ahrne 1994). How organizations come into existence "and how they evolve are fundamentally influenced by the institutional framework. In turn they influence how the institutional framework evolves" (North 1990, 5). There is thus a mutual reinforcement between organizations and institutions. What, then, are institutions?

According to the neoinstitutionalist turn in political science during the last decade, institutions are certain social phenomena that can create stable patterns of collective and individual behavior (Premfors 2001; see also Peters 1999). Rules, procedures, and certain structures can constrain and/or facilitate actors' behavior, but they can also form actors' preferences and interests. There is thus a wide array of social phenomena that can be called institutions, and they can have different effects on collective and individual behavior. In the sociological institutionalism approach, institutions take on a rulelike status in social thought and action (Meyer and Rowan 1977/91; March and Olsen 1989; DiMaggio and Powell 1991). Other approaches such as rational-choice institutionalism and historical institutionalism, do not require a taken-for-granted status to define something as an institution (Shepsle 1989; North 1990; Thelen and Steinmo 1992). In the political world there are very few institutions that can be regarded as taken for granted in the sense that they are not contested. However, there are some rules that are less contested than others, and these, therefore, to a large extent, influence individual and collective behavior. How these rules become more or less taken for granted is a process and it can vary over time. We should therefore search for processes of institutionalization instead of trying to identify fixed static institutions.

The question of how organizations are institutionalized can be studied as the emergence of an organizational field. Drawing on Bourdieu's work on fields, a literature on organizational fields has developed that emphasizes how organizations are dependent on each other, but also that they simultaneously compete with each other (DiMaggio 1983; DiMaggio and Powell 1991; Scott 1995). Bourdieu uses the term *champ* to convey competition and struggles between participants in a field (1996). A field is a system of objective relations between positions occupied by agents and institutions that struggle for something held in common by them (Broady 1998). A field signifies "both common purpose and an arena of strategy and conflict" (DiMaggio 1983, 149). In the art field, for instance, the struggle is about the definition of valuable art and the authority to make authoritative statements about artistic values (Bourdieu 1996). One definition of an organizational field is "sets of organizations that together accomplish some task in which a researcher is interested" (DiMaggio 1983, 148). By using the concept of an organizational field, it is thus possible to include in the analysis all organizations operating in the same domain.[2] A more elaborated definition is that an organizational field consists of organizations "that in the aggregate constitute a recognized arena of institutional life: key suppliers, resource and product consumers, regulated agencies, and other organizations that produce similar services or products" (DiMaggio and Powell 1991, 64-65; Scott 1998). This definition leaves out an important methodological question, namely, whether the field boundaries are drawn by the analyst (objectively defined) or by the actors (subjectively defined). The problem of how to define the field boundaries is crucial in Bourdieu's work on fields (1996). He argues that a field possesses a well-developed autonomy in relation to other fields and to the general external environment (Bourdieu 1996; Broady 1998). However, this criterion is seldom fulfilled in empirical studies (Broady 1998). It is difficult for the analyst to identify the autonomy since fields that are studied are being transformed or are in the making and will therefore always be characterized by weak autonomy. It is also difficult for the actors to define the field because the field boundaries are at stake themselves. Participants in a field try to differentiate themselves "from their closest rivals in order to reduce competition and to establish a monopoly over a particular subsector of the field" (Bourdieu and Wacquant 1992, 100). This constant struggle between participants in a field suggests that there is an ongoing reproduction of the structure of the field and that the field is a work in motion. The complex picture of the formation of the field boundaries means that it is extremely difficult to choose between either an objective or a subjective definition. It is necessary to combine these two methods of defining the boundaries of the organizational field (see below).

According to Bourdieu, the participants in the field are loosely connected to each other. To think in terms of fields is to think relationally, but these relations are objective and they exist independently of individual consciousness (Bourdieu and Wacquant 1992). The relations must not be interpreted as interactions (Bourdieu and Wacquant 1992). "In analytical terms, a field may be defined as a network, or a configuration, of objective relations between positions. These positions are objectively defined, in their existence and in the determinations they impose upon their occupants, agents or institutions, by their present and potential situation (*situs*) in the structure of the distribution of species of power (or capital) whose possession commands access to the specific profits that are at stake in the field, as well as by their objective relation to other positions (domination, subordination, homology, etcetera)" (Bourdieu and Wacquant 1992, 97). The participants are not held together other than by the ambition to be authoritative interpreters of art, literature, or fashion. Scott and Meyer argue that the organizational field connotes "the existence of a community of organizations that partakes of a common meaning system and whose participants interact more frequently and fatefully with one another than with actors outside of the field" (Scott and Meyer 1994, 207-8). Thus, according to Scott and Meyer, organizational fields are bound to each other in ideational and relational terms. DiMaggio argues that one process toward making an organizational field is the development at the cultural level "of an ideology of the field" (DiMaggio 1983). The formation of an organizational field will thus result in a sense of belonging among the participants.

In my view, the questions, What kind of interactions are there between the organizations? and How do the organizations critically influence one another's performance? are empirical ones. As DiMaggio puts it, "the extent to which a field constitutes a network of interaction is always an empirical question" (DiMaggio 1983, 149). Does the interaction vary over time within the field? What are the factors that influence the degree of interaction? It is also possible that the organizations monitor each other but that they do not develop any patterns of interaction (Forssell 1992). Furthermore, are the organizations linked by cultural and ideational bonds, or are we dealing with interorganizational activity that just happen to exist in the same policy area? Will the interactions between the organizations lead to an embeddedness that could result in a sense of belonging? The question of the character of the interactions between organizations in an organizational field is thus an empirical one.

Another empirical question is: What kinds of institutional glue hold the organizations together? I argue that organizational fields consist of various rules of the game that are institutionalized and to a large extent taken for granted. Each organizational field has its own images of the external environment, its own core of ideas about what constitutes a threat, how a problem is diagnosed, what counts as a problem in the first place, and how to approach this perceived problem. The term *institutionalized organizations* refers to organizations that internalize certain rules and accomplish what is expected of them in order to be politically legitimate (Meyer and Rowan 1977/91). In my usage and definition of organizational field I refer not only to interaction between organizations that together accomplish some task in which a researcher is interested, but also to the fact that the organizational field must entail institutionalized rules.

In this book, organizational fields are defined as consisting of inter- actions and relationships between organizations that together accomplish some task in which a researcher is interested, with these interactions be- ing held together by common rules of the game. What are the rules of the game? They are the basis of the field in question, from which the external environment is interpreted and from which political life is organized and handled. To identify the rules of the game and the ideas of the organiza- tional field, I make a distinction between regulative rules—legal and other concrete rules[3]—and meta and constitutive rules which "define the set of practices that make up a particular class of consciously organized social activity—that is to say, they specify what counts as that activity" (Ruggie 1998b, 871; see also Rawls 1955; Searle 1994; Ruggie 1998a). This kind of identity meta rule-making can thus be characterized as the construction of constitutive rules without which the regulative rules cannot take effect.

The question of what kind of regulative and constitutive rules there are in various organizational fields is an empirical one. So is the question of how coherent these rules are within an organizational field. This is especially interesting since in this empirical analysis we are dealing with the European Commission and other complex organizations. These multiorganizations are characterized by ambiguity, different policy styles, multiple interests, identities, functions, inconsistent organizational setup, and vague organizational boundaries (March 1994). How can we empirically study these rules? Regulative rules are rather easily identified since they concern legal rules, for instance, the treaties and other legal rules. By constitutive rules I mean rules that, in a fundamental way, determine how an issue is to be interpreted. Such rules are, of course, not easily separated from regulative rules, but various legal articles in the EU treaties are based on different ideas on the EU as an organization for political cooperation. These ideational rules can be descriptive and/or prescriptive. The EU, for instance, can be described by actors in terms of supranational or intergovernmental cooperation, but it can also be argued that the EU should be more supranational or be based on intergovernmentalism.

Hence, different rules—regulative and constitutive—in the fields determine how an issue is conceptualized and defined. Actors are thus following rules, but actors can also influence the rules of the game. An organizational field takes shape in a process in which actors, based on the regulative and constitutive rules, determine what counts as activity. The two rationalities and logics—the logic of consequentialism and the logic of appropriateness—are relevant in the study of the dynamics in European organizing (cf. Green Cowles et al. 2001). Thus, "actors both calculate consequences and follow rules" (Laegreid and Roness 1999, 308; see also March and Olsen 1998; Marcusson et al. 1999; Fierke and Wiener 1999).

How an issue is labeled is, in this study, described with reference to frames. The term frame was initially introduced by Goffman to denote that "when an individual in our Western society recognizes a particular event, he tends, whatever else he does, to imply in his response (and in effect employ) one or more frameworks or schemata of interpretation of a kind that I call primary . . . a primary framework is one that is seen as rendering what would otherwise be a meaningless aspect of the scene into something that is meaningful" (1974, 21). The more politically oriented activity of framing can be seen as two types of activity—"diagnostic framing" and "prognostic framing" (Snow and Benford 1988). The first type of framing concerns the identification of the problem/issue at hand and why this problem has occurred, for example, the attribution of blame or causality. This activity can also be referred to as the agenda-setting phase in the policy-making process. The second type of framing is more action oriented and is thus focused on solutions to the problem and the suggestion of strategies. This usage of the frame concept is also evident in the definition by Schön and Rein, who argue that the activity of framing consists of struggles "over the naming and framing of a policy situation" and that these struggles "are symbolic contests over the social meaning of an issue domain, where meaning implies not only what is at issue but what is to be done" (Schön and Rein 1994, 29; see also Gamson 1988; Jachtenfuchs 1996).

The interesting question is why one particular frame appears rather than another (cf. Foucault 1972). I argue that this depends on the organizational field within which the issue is framed. Depending on the rules that hold the organizational field together, issues are framed in certain ways. This means that actors who frame an issue are part of an organizational field that consists of regulative and constitutive rules that provide stability and meaning to social behavior (cf. Meyer and Rowan 1977/91; Scott 1995). Frames can thus construct identities since they are based not only on regulative rules but also on constitutive rules in the field. "The environment in which actors operate is given meaning through ongoing processes of social construction. This means that there is an inherent connection between the social construction of the external environment and the interests that actors acquire" (Rosamond 1999, 658). The actors interpret each other's moves and "constantly renegotiate the reality in which they operate" (Kratochwil 1989, 101). Frames can also be seen as instruments for pursuing various interests. Globalization and other catchwords, for instance, can be used by actors to promote a specific set of policy solutions to various external threats (Rosamond 1999). Actors are not merely finding circumstances but are also very much involved in the making of circumstances (Ruggie 1998b).

I argue that both the power and identity dimensions of framing activities between the fields are important in the empirical analysis. In this study of the emergence of an organizational field on armaments, this means that actors who strive for influence on how to organize cooperation are simultaneously part of institutions that, notes W. Richard Scott, consist "of cognitive, normative, and regulative structures and activities that provide stability and meaning to social behavior" (in Peters 1999, 106). The power dimension is important in order to study the conflicts between various actors that pursue different frames. Hence, we are interested in answering the question: What is at stake? The identity dimension of the analysis of frames is essential to the study of how an issue is interpreted and how actors within a field organize collective experience (Snow and Benford 1988). What are the underlying constitutive rules behind a certain frame?

Having discussed how organizational fields are created and how they set the rules of the game, how can we analyze and explain what is occurring when organizational fields change, transforming into new organizational fields? The rules of the game are occasionally contested and challenged. What are the dynamics behind changes in organizational fields? The literature on organizational fields argues that the structuration of a field means that the organizations begin to resemble each other, which is called institutional isomorphism (Meyer and Rowan 1977/91). Once again we are dealing with an institutional concept from sociological institutionalism and its emphasis on social interaction and rules that are taken for granted. Three different ideal types of how this taken-for-grantedness emerges are identified in various studies: coercive isomorphism, mimetic isomorphism, and normative isomorphism (Meyer and Rowan 1977/91). Since they are ideal types, they cannot be easily identified in the empirical analysis. The borders between them are seldom clear-cut (Stern 1999). Coercive isomorphism is a process whereby the organizations begin to resemble each other due to legislation and other coercive rules. Thus, legislation forces the organizations to follow the same rules and places limits on their room to maneuver. Mimetic isomorphism causes resemblance when the organizations imitate each other. This is especially the case in a field that is perceived as uncertain and turbulent. Indeed, uncertainty is a powerful force that encourages imitation (DiMaggio and Powell 1991). One example of mimetic isomorphism is the European and American effort to organize their own industries according to the Japanese model (DiMaggio and Powell 1991; Stern 1999). Normative isomorphism is a social process that creates rectification within an organizational field primarily by the professionalization of society. People employed in various organizations tend to have similar educational backgrounds, norm systems, and ideas about how to create good organizations (Stern 1999).

These three ideal types tell us very little about the specific ways in which the organizations resemble each other; they only identify the general *mechanisms* behind the changes—coercive or voluntary. Organizations can move closer to each other in several ways, and determining if and in what ways they tend to resemble each other is an empirical issue. The organizational setup and the decision-making structure can become more similar, for instance, with regard to what kind of units, procedures, etc., exist in the organization. Organizations can also begin to resemble each other in ideational and cultural terms, for instance, in terms of how issues are identified, perceived, and handled. The focus in this study is on the latter case. In what ways does the emergence of an organizational field of armaments require the participants in the field to begin to frame the issue of armaments in new ways, and how should we interpret these changes? Sahlin-Andersson argues that it is rather common that organizing activities are a result of a convergence of actors' expectations and interests (Sahlin-Andersson 1986; see also Gibson and Goodin 1999). They act together, but from different institutional perspectives. The aim of the organizing process is unclear, and thus the actors can use the cooperation in ways that are most important in terms of their own interests and positions (Sahlin-Andersson 1986). Sahlin-Andersson raises a fundamental question in the study of changes in organizational fields. Can changes in how an issue is framed be interpreted as changes in the rules of the game, or are the changes only the result of the actors involved having reached a lowest common denominator? In this study, changes are characterized as thin if they concern only regulative rules, and as thick if they also concern constitutive rules.

In the literature on fields it is argued that one important factor behind turbulence within a field is when there is a struggle between the occupants of the dominant positions in the field and the pretendents. The dominants' strategies are tied to continuity, whereas the pretenders are interested in discontinuity and rupture (Bourdieu and Wacquant 1992). "Those who dominate in a given field are in a position to make it function to their advantage but they must always contend with the resistance, the claims, the contention, 'political' or otherwise, of the dominated" (Bourdieu and Wacquant 1992, 102). Thus, an important factor behind changes in the field is when the existing rules of the game are called into question and when they are challenged. The emergence of a new organizational field can, however, be difficult to separate from normal dynamics in fields (see Bourdieu 1996). It is certain there will be disapproval from actors on various authoritative ways of defining fine art or literature (Bourdieu

1996).

The crucial precondition for changes in the field is not disapproval of the existing rules of the game, but that the rules must be seriously challenged. Since organizational fields consist of institutionalized rules, we should look for changes in these rules, especially changes of constitutive rules. At the same time, I am aware that organizational fields are constantly reproduced and that they evolve through stages of increasing structuration. We should therefore focus on the process of institutionalization and try to identify important change indicators in that process of consolidation (cf. Giddens 1979; 1984; see also Ekengren 1998). "In all forms of interaction certain patterns gradually emerge—certain logics of interpretations, accepted conceptions and experiences. All interaction may therefore be regarded as a way of organizing as structuring. Structures are, in other words, not given but are formed and reformed through activities" (Sahlin-Andersson 1986, 66; 1998). Drawing from Giddens's work, the structuration process of an organizational field can be studied as the result of five components: "an increase in the level of interaction among organizations in the field, an increase in the load of information on organizations in a field, the emergence of a structure of domination, the emergence of a pattern of coalition; and the development, at the cultural level, of an ideology of the field" (DiMaggio 1983, 150). This means that in order to study the emergence of an organizational field it is necessary to analyze both the quantity and quality of the interactions and contacts between organizations, the power relations, and whether organizations develop institutional rules.

The Empirical Study

The analysis of European cooperation on armaments is based on three theoretical and analytical concepts: organizations, organizational fields, and frames. The analysis focuses on how various types of organizations, formal and informal, take part in two organizational fields—defense and market (chapters 3 and 4)—and how these two fields move closer to each other (chapter 5).

The organizational fields and frames are deduced from the empirical material, interviews, and official statements. A market field/frame on the issue of armaments concerns the EU's industrial development, research and technological development (RTD), and other policies within the community pillar whereas the defense field/frame concerns the EU's common foreign and security policy (CFSP). I have interpreted the autonomy of the fields based on the empirical material. Thus, how the borders of the field are drawn and how its relations to other fields are specified are dependent on how the analyst constructs her research object and how the empirical material is interpreted. My ambition with the analysis of the formation of a common organizational field on armaments is to identify the process of consolidation between the market and defense fields. Based on DiMaggio's (1983) work on organizational fields, I study the transformation of the defense and market fields into a common organizational field on armaments as consisting of three consolidating processes:

1. Reframing of the issue of armaments
2. New forms of formal/informal domination and authority structures
3. Scope and depth of organizational interaction.

The relationship between these processes is not necessarily causal although it is likely that increased activity and interaction between organizations from different fields will lead to a common frame on the issue of armaments. It can also be the case that the issue of armaments is first reframed within one field and that this new understanding of armaments will lead to increased activity and interaction between the two fields. Furthermore, new types of organizing activities can challenge the existing authority structures. These activities will lead not only to new forms of domination, but also to a reframing of the issue of armaments and to more interactions. An empirical question is to what extent the reframing is a result of changes of regulative or constitutive rules in the two fields. Are we dealing with profound changes in the fields, or is a reframing only the result of pragmatic considerations by the actors involved? Another question, studied in chapter 5, asks what these processes of consolidation mean in terms of institutional isomorphism. Do the organizations in the field in the making begin to resemble each other, and, if so, what are the general mechanisms behind this isomorphism? Are the organizations in the fields legally forced to reframe the issue of armaments so that it must contain both a market and a defense frame? In what ways is the threat from the United States a driving force behind a European effort to strengthen its cooperation on armaments?

An important theme in this book is that certain issues seem to circulate most of the time in the European integration process. They travel both in space and in time. We begin our empirical journey on the issue of armaments in an organizational field that is occupied with European foreign and security policy. We then move on to see how the issue of armaments has been part of the EU's civilian-driven project.

Notes

1. The concept of governance is developed and discussed in both international relations and comparative politics (Rosenau and Czempiel 1992; Rosenau 1992; Rhodes 1997; Pierre 2000). The overall discussion in these subdisciplines concerns how "authoritative allocation takes place without or outside government" (Eising and Kohler-Koch 1999, 3-4). The definitions of governance vary. Rhodes argues that governance is synonymous with a new process of governing (1997). Governance can also entail different modes of governing patterns (Eising and Kohler-Koch 1999). The notion of governing without governments is one of the fundamental starting points in the field of international relations and in the study of international regimes (Ruggie 1975; Krasner 1983).

2. A concept related to organizational field is policy domain (Fligstein and McNichol 1998). Fligstein and McNichol define a policy domain as a group of actors who are organized to "participate in a collective debate with the goal of affecting the content of legislation or agreements. These actors include government organizations and might include organized interest groups, some of which might be other government organizations" (Fligstein and McNichol 1998, 61). Thus, policy domains are "broad sets of issue arenas that both state and non-state actors consider" (Fligstein and McNichol 1998, 61). It is also obvious that the regime concept is related to the concept of organizational fields, that is, how certain ideas and concepts constitute international authority and the game nations play (Adler 1997). The regime analysis also recognizes competition. Existing institutional arrangements can be challenged (Aggarwal 1998). Another concept that is related to the organizational field concept is that of networks (Sahlin-Andersson 1986). Indeed, there are many concepts that are related to the concept of organizational fields, which in itself can be interpreted in multiple ways. The overall implication in these concepts is the notion that a complex political process requires a concept that catches multiple actors, issues, and processes (Sabatier 2000). It is my understanding that the regime literature is more consensus oriented and that the literature on organizational fields is more focused on conflicts and interdependence (Hasenclever et al. 1997). This is also the case with the literature on policy domains. In addition, the notion of a policy domain often implies that the actors concerned have fixed interests in the policy area, and that they pursue different goals (Héritier 1999). The interactions within the domain thus consist of exchanges and bargains over various resources to reach a policy decision

(Lauman and Knoke 1987). Although I believe that actors have different interests, I am also interested in how interests change and how they are formed and influenced by the organizational field.

3. In my usage of the concept of regulative rules I would include both coercive rules and rules based on soft law.

Chapter 3
The Defense Field

*In Kosovo, we have all come face to face with the European future, and
it's frightening. . . . The lesson is clear. Europe should do better.*
—George Robertson, British Defense Minister,
8 September 1999

*European integration goes beyond agriculture and economics. It aims at
creating a political union which is underpinned by a credible political
and military capability to act.*
—Rudolf Scharping, German Defense Minister,
11 January 2000

This chapter focuses on how the issue of armaments is framed as part of
a defense organizational field. The first section of the chapter presents the
main steps in the development of Europe's security and defense policy
since the early days of the European integration process. This section thus
outlines the components of the defense organizational field from which the
issue of armaments has been framed. The organizations in this field, and
how they have framed and handled the issue of armaments, are analyzed
in the second section of the chapter.

Europe's Security and Defense Policy

Security and defense issues have, since the end of the Second World War, been an important part of European cooperation. The mutual defense agreement between the United Kingdom and France concluded in 1947 was, one year later, extended to include the Benelux countries and became the Brussels Treaty (Bretherton and Vogler 1999). "The members of the Brussels Treaty Organization agreed 'to take such steps as may be held necessary in the event of renewal by Germany of a policy of aggression.' This agreement was reinforced by a generalized collective defense commitment should any of the signatories be 'the object of an armed attack' (Article V)" (Bretherton and Vogler 1999, 199). Rather soon, however, it was clear that peace was threatened by the rapidly deteriorating relations between Western governments and the Soviet Union. The signatories of the Brussels Treaty Organization therefore began negotiations with the U.S. and Canadian governments "for establishment of a transatlantic collective defense arrangement" (Bretherton and Vogler 1999, 199). In August 1949 the so-called Washington Treaty—which created the North Atlantic Treaty Organization (NATO)—entered into force. Even if the creation of NATO solved many of the security problems for the nations of Western Europe, there were still questions left unresolved. It did not "resolve the internal security problems of Western Europe, of which the most sensitive was the relationship between France and Germany. Nor did it adequately determine the role to be played by the West Europeans in their own defense" (Bretherton and Vogler 1999, 200; Forster and Wallace 2000).

European efforts to create a more independent security policy have not been absent, although security and defense issues were not part the three founding treaties, the European Economic Community (EEC), Euratom, and the Treaty of Rome. In October 1950 the French government put forward the proposal for the creation of a European Defense Community (EDC) among the six members of the European Coal and Steel Community (ECSC). An important reason behind the French initiative was a desire to reduce the risks related to the existence of a militarily strong Germany (Grosser 1980; Sloan 1985; Rees 1998). The proposal—known as the Pleven Plan—gave rise to the suggestion that a federal army should be under the control of a supranational authority (that is, an elected EDC Assembly). At a NATO council meeting in Lisbon in 1952, a conference report on EDC reaffirmed the purpose of the plan[1]: "To create a European Defense Community which can fulfill the imperative requirements of military effectiveness; to give the Western World a guarantee against the rebirth of conflicts which have divided it in the past; and to give an impetus to the achievement of a closer association between the Member countries on a federal or confederal basis" (Sloan 1985, 17-18).

The Pleven Plan and the political vision of an EDC thus consisted of several supranational components. "There was to be a dedicated European Minister of Defense, answerable to a political authority of a Council of Ministers and an Assembly. They were to exercise control over a force of some 100,000 personnel, effectively a European army. This was to be financed out of a common budget and the Minister would have been accorded responsibility for raising and equipping forces from the member states through a common procurement system (Rees 1998,7). Indeed, the concluding plan "would have transformed the Six into an effective federation, with a European executive accountable to a directly elected European Parliament" (Forster and Wallace 2000: 463).

The EDC Treaty was signed in May 1952, with the blessing of the United States, by Germany, Belgium, Holland, and Luxembourg (Rees 1998). Italy and France were lagging behind in the process (Rees 1998). "In France, there was a delay before submitting the Treaty to the Assemblée Nationale and as time dragged on, there were growing doubts as to whether it would be passed" (Rees 1998: 8). There would be at least three problems if the EDC were to fail. First, disagreement in Europe could indicate to the Soviets that the Europeans were not willing to contribute to their own defense. Second, without the EDC there was no institutional context in which Germany could rearm. Third, a lack of unity in Europe might mean that the United States would be less willing to support European defense. In addition, already in 1951 the British government had decided not to participate in the EDC. This was surprising since Sir Winston Churchill had, in August 1950, called for the creation of a European army under the authority of the European defense ministers and subject to democratic control by the Assembly of the Council of Europe (Bainbridge 1997). In August 1954 the EDC Treaty was rejected by the French National Assembly. One important motivation behind the French rejection was doubts about the supranational components in the agreement. This could limit "France's sovereignty and it was feared that this might undermine the country's sense of national identity" (Rees 1998, 8). An abolition of national military academies was shocking to the French public, and among the political elite there were worries over the possibility of controlling Germany (Grosser 1980). There was also a general sense of

France's exposed situation without any firm commitments for the future from either Britain or the United States (Rees 1998, 8). Due to the death of Stalin and the Korean armistice, the threat perceptions among the political elite changed. "Although the French remained concerned about the potential threat, they had become more relaxed about the Russian threat and somewhat more wary of American attentions" (Sloan 1985, 25; see also Forster and Wallace 2000).

As a result of the failure to include a defense component in the EU, the Brussels Pact was transformed into the Western European Union (WEU) in 1955, and Germany joined NATO as an alternative to the EDC. In the European context, the WEU can be seen as the replacement for the EDC. Defense issues continued to be discussed among Western European countries, but not within the EU. This happened to some extent within the WEU, but the crucial security organization in Europe for several decades was NATO. The search for greater political unity among the EU member governments continued, and in 1969 the heads of state and government of the original EU six "instructed their Foreign Ministers 'to study the best way of achieving progress in the matter of political unification, within the context of enlargements' of the European Community" (Sloan 1985, 50). This was the beginning of the process of european political cooperation (EPC), which aimed to ensure regular exchanges of information and consultation among the EU member governments on important international problems and "to strengthen their solidarity by promoting the harmonization of their views and the coordination of their positions and where it appears possible and desirable, common actions" (Sloan 1985, 50). EPC received legal status in the Single European Act (SEA) in 1987 and was restructured as the common foreign and security policy in the Maastricht Treaty (see below). Although the EPC was established to deal with foreign policy issues, the member governments expanded the scope and content of their consultation in a way that put the EU "into the gray area of 'security policy'" (Sloan 1985, 51). Indeed, the border between defense and foreign policy issues is diffuse, which is illustrated by various political initiatives during the 1960s and onward. According to the so-called French Fouchet Plan in 1961, new policy areas would be incorporated into European cooperation, most notably in the areas of foreign policy and defense. The plan aimed at adopting a common foreign and defense policy (Grosser 1980). The cooperation was to be intergovernmental and based on unanimity (Grosser 1980). This plan failed for several reasons. This failure once again illustrates the bewildering complexity of European cooperation in the defense area, complicated by such issues as the role of the United States, or, to put it differently, the balance between NATO and the EU; the supranational/intergovernmental conflict dimension in the EU, and whether the United Kingdom should be permitted to participate.

The 1960s can be broadly characterized as a decade of Cold War whereas the 1970s can be labeled a period of détente and Ostpolitik. As the 1970s ended, however, tensions once again rose between East and West. In 1979 NATO decided to respond to a perceived Soviet buildup of medium-range nuclear missiles with a two-track decision: while NATO was prepared to negotiate concerning the dismantling of the Soviets weapons, it at the same time retained the option to deploy its own medium-range missiles if negotiations failed. During this period of NATO domination of European defense policy there was one new plan introduced within the EU: the Genscher-Colombo Plan in 1981, which aimed to add a security and defense component to the existing EPC (Missiroli 2000).[2] Although defense issues were not formally part of the competence of the EPC, these issues were in practice already dealt with by the EPC (Sloan 1985). So, the Genscher-Colombo Plan would to some extent only codify what was already under way. "The members of the European Community have been reticent, however, about formally acknowledging the role of EPC in security policy" (Sloan 1985, 180). One reason for this European reluctance was the fear of "trespassing in territory belonging to NATO" (Sloan 1985, 180). The Genscher-Colombo Plan was not adopted and resulted instead in a declaration in 1983 that watered down the role of security policy in the EPC (Sloan 1985, 180).[3]

The failure of the Genscher-Colombo Plan has been interpreted as an opportunity for a more active role for the WEU (Sloan 1985). In 1984, the WEU Council was reactivated by the seven member states that were also members of the EU, and it was agreed that defense and foreign ministers would meet on a regular basis (Missiroli 1999). In 1987, the WEU Ministerial Council adopted the Hague Platform (Missiroli 1999). In this way the WEU functioned as a forum for defense issues, but outside the EU framework and without a U.S. presence. "In other words although defense was still a taboo, further European integration was being increasingly associated with the idea of a defense dimension, if still outside of the EC/EPC framework" (Missiroli 1999, 22-23). The political and constitutional breakthrough for defense issues in the EU came in Maastricht in 1991 when the heads of state and government of the member states of the European Union decided that the Treaty on European Union should establish a common foreign and security policy. The treaty contained the often-quoted formulation of the "eventual framing" of a common defense policy that "might in turn" lead to a common defense (Article J.4.1). TEU thus did not promise a European defense immediately, but it opened the way for such developments in the future. Furthermore, it was also a breakthrough for more coordinated activities among the EU, the WEU, and NATO. Until then, the three organizations had dealt with security issues in their own separate ways. "The European Union restricted itself to basic consultation on foreign policy issues and upheld its decade-long taboo on military affairs; the recently reactivated Western European Union was only starting to come into its own; and NATO was only just commencing its own process of adjustment to the new strategic environment" (De Spiegeleire 1999, 59, see also Rees 1998). A WEU declaration, which was attached to the Maastricht Treaty, declared that the "WEU will be developed as the defense component of the European Union and as the means to strengthen the European pillar of the Atlantic Alliance. To this end, it will formulate a common European defense policy and carry forwards its concrete implementation through the further development of its own operational role" (Maastricht Treaty, 10 December, 1991). The most important component of that operational role was the Petersberg Declaration of June 1992. The "Petersberg tasks" include humanitarian and rescue tasks, peacekeeping, and crisis management. Various multinational forces created during the 1990s are important steps toward a European military capability (for instance, the Eurocorps, Eurofor, Euromarfor—for an overview see CEPS 1999). The decision to establish the Petersberg tasks was not followed by new military forces. "De facto, however, WEU was given no military forces under its direct command and remained dependent on NATO for surveillance, intelligence gathering and long-range transport support" (Missiroli 1999, 25; Sjursen 1998).

The question of the role of the United States and NATO in the European security and defense identity is a recurrent theme during the 1990s (and, as we have seen, during the whole period after the Second World War). The relationship between the WEU and NATO has, however, improved. In April 1999 NATO presented a new strategic concept of which one component was a new pragmatic and flexible view on the relationship between NATO and the WEU (De Spiegeleire 1999). Indeed, NATO's organizational structure has been Europeanized in recent years, and the WEU's organization has been made more compatible with NATO—especially due to the Petersberg tasks (De Spiegeleire

1999). Concrete measures have also been taken to enhance the operational contacts between the two organizations (De Spiegeleire 1999).

In Amsterdam in 1997 the European defense dimension was strengthened (Article J.7—now Article 17 of the consolidated version of the Treaty on European Union); "The common foreign and security policy shall include all questions relating to the security of the Union, including the progressive framing of a common defense policy . . . which might lead to a common defense, should the European Council so decide" (Article 17.1). The WEU is, according to the treaty, "an integral part of the development of the Union, providing the Union with access to an operational capacity" (The EU Council 10 May 1999). Thus, as in the Maastricht Treaty, there was a call for enhanced cooperation between the two organizations. However, most of the difficult issues that were left unresolved by the Maastricht Treaty remained after the Amsterdam meeting (Sjursen 1998).

The conflict in Kosovo during 1998 and 1999, combined with a new British and French military initiative in 1998, started a discussion on the need for a European common defense policy and military capacity. Hence, the war in Kosovo seems to have further legitimated the process towards filling the concept of a European security and defense policy with material contents, that is, a common defense policy and military capacity. There was also a German determination to play a more active role within European security issues, and the so-called Saint Malo agreement between France and the United Kingdom was extended to include Germany (*Financial Times* 29 May 1999). The changed policy of the British government is also identified as an important factor in the EU's agenda-setting process on defense issues (Missiroli 2000). In the Saint Malo Declaration between British Prime Minister Tony Blair and French President Jacques Chirac and French Premier Lionel Jospin, the three leaders agreed that the "Union must have the capacity for autonomous action, backed up by credible military forces, the means to decide to use them and a readiness to do so, in order to respond to international crises" ("Joint declaration" 1998). Thus, they left open the option of European military action outside NATO. At the Franco–German summit in May 1999 (in the so-called Toulouse Declaration) it was declared that there was a "determination to contribute all their weight so that the EU equips itself with the necessary autonomous means to decide and deal with crises" (*Financial Times* 31 May 1999).[4] In a speech given on 30 May 2000, President Jacques Chirac noted that there had been "considerable progress during the Finnish and Portuguese presidencies of the EU toward the realization of the goals defined by Great

Britain and France during the Saint Malo agreement" (*Financial Times* 31 May 2000).

In June 1999 the so-called Cologne Declaration was presented. It is a detailed blueprint for strengthening the Union's ability to launch military action. The declaration "mapped out how the Union could take independent military action in future, without relying on the political support of other NATO countries or the alliance's equipment" (Simon Taylor 1999). "We, the members of the European Council, are resolved that the European Union shall play its full role on the international stage. To that end, we intend to give the European Union the necessary means and capabilities to assume its responsibilities regarding a common European policy on security and defense" (Declaration of the European Council on strengthening the common European policy on security and defense, paragraph 1). Furthermore, it was stated that the WEU could be integrated into the EU. "The Treaty also provides for the possibility of integrating the WEU into the EU, should the European Council so decide" (Presidency Report on Strengthening of the common European policy on security and defense, paragraph 1). On 15 November the EU's High Representative for Common Foreign and Security Policy, Javier Solana, was chosen as the new Secretary General of the Western European Union. This decision "set the seal on a decision to merge all of the WEU's operational and planning capacities into the EU, giving the Union its first real option for military action in history" (*European Voice* 18 November 1999). This means that only the collective security guarantee (Article 5) of the WEU and the WEU Assembly will remain outside the new construction.

By October 1999, the former Secretary General of NATO, Javier Solana, already was formally established as the EU's High Representative for Common Foreign and Security Policy. He will be assisted by an early warning and planning unit. Furthermore, he is expected to chair the new political and security committee (an equivalent to the North Atlantic Council [NAC] within NATO). A new military committee will also be established that will be subordinated to the political and security committee. The new body, consisting of senior-level officials from national foreign and defense ministries, began its work in an advisory capacity as early as January 2000. This new decision-making structure also includes, on top of the permanent political and security committee, a military committee and a body of military staff. There will also be a special civil crisis management committee that was decided in May 2000 and confirmed at the Feira European Council meeting in June 2000. At the summit in Helsinki, the European Council agreed to create a new security and defense capability that would allow the EU to take military action in crises should NATO decide that it did not want to spearhead a campaign. "The European Council underlines its determination to develop an autonomous capacity to take decisions and, where NATO as a whole is not engaged to launch and conduct EU-led military operations in response to international crises. This process will avoid unnecessary duplication and does not imply the creation of a European army" (Helsinki European Council, Presidency Conclusions 1999). The Council decided to set up an EU rapid-response force of up to 60,000 troops that could be deployed at sixty days notice and could stay on the ground for at least one year (Presidency Conclusions Helsinki European Council). Furthermore, the Presidency Conclusions from Helsinki also stated that the defense ministers would be involved in the common European security and defense policy; when the General Affairs Council discussed matters related to the CESDP, "Defense Ministers as appropriate will participate to provide guidance on defense matters." Thus, the defense ministers did not constitute a council of their own in 1999, but they were part of the General Affairs Council every six months; that is, they worked together with the foreign ministers. Informal meetings did, however, take place between the EU's defense ministers from the autumn of 1998.

To sum up, the political process within the EU has recently moved from a situation in which the EU did not have the political will or the institutional or organizational capacity for dealing with defense issues, to a situation in which defense issues are not only on the political agenda but also the object of a new institutional and organizational setup. The consolidation process has so far entailed internal organizing processes within the EU and NATO, for instance between parts of the Commission that deal with CFSP issues and the High Representative within the Council, and between the EU and NATO—the authority relations between them and concrete organizational linkages in a situation in which most functions of the WEU will be incorporated into the EU. These organizing processes are likely to continue. This will also mean that new types of organizational and institutional ties will emerge between these two organizations. The implications of a European defense identity, that is, a European planning, policy, and capacity dependent on NATO or a more independent European planning, are an example of this.

The similarities between the EDC and the present process are striking. One similarity is, of course, the perceived need for a European defense capacity and policy. This time, however, this is clear not only on the level of rhetoric but also in practice. Another resemblance is the common decision-making structure that was part of the EDC plan and that is realized today. However, an important difference is that the EDC had clear federalist features whereas the current plans are intergovernmental. The big difference between the 1950s and the situation during the 1990s and early years of the millennium is the end of the Cold War. The role of the United States in defense in Europe was more direct during the Cold War than has been the case since 1990. In the 1950s in particular, the United States was involved in discussions with British political leaders, and to some extent with Gaullist France, about creating a directorate comprising only the most important members of NATO. There has been no such discussion between the U.S. government and individual governments in Europe concerning security affairs after 1990.

In political declarations the linkage to the EDC and the present process is seldom underlined. One exception is a statement by the French Prime Minister. Lionel Jospin has explicitly mentioned the Cologne process and the decision in Helsinki in relation to the failure of the EDC in 1954. In a statement to the French National Assembly on 9 May 2000, he noted, "I would ask you to consider how far we have come since that date. For the past few months France has been performing a vital role in creating credible prospects for European defense." The historical defining moments of the summits in Cologne and Helsinki are also outlined by the German Defense Minister, Rudolf Scharping, but there is no explicit reference to the EDC: "It took major steps to establish new political and military structures and develop more effectively military capabilities for the full range of conflict prevention and crisis management tasks defined in the EU Treaty" (Scharping 2000a). In another speech he underlines the historical difficulties for the Europeans with respect to their efforts to build a defense capacity, and says, "Today, we have every chance of bringing about a change. Of modernizing Europe's armed forces" (Scharping 2000b). The British Defense Minister has clearly underlined that the so-called Headline Goal, decided at the Helsinki European Council in December 1999 (to deploy rapidly and sustain 60,000 troops), does not mean a European army. "We should be clear: this is not a European army. Member States have not committed themselves to a standing military formation, but to the ability to put together the right mix of forces and capabilities to suit particular circumstances" (Hoon 2000).

The EU, NATO, WEU: WEAG/WEAO and the Defense Frame on the Issue of Armaments

The European Union

Attempts to harmonize armaments procurement in Europe have a long history (Hayward 1997; Cornish 1997). Until the mid–1960s, the political initiatives came primarily through NATO. A European voice was established in the mid–1970s by establishing IEPG (the Independent European Programme Group [IEPG]), the predecessor to WEAG (see below). So, the lack of activities within the European Union on the issue of armaments has not prevented European activities outside EU institutions in the CFSP political process, that is the second pillar of the European Union. In the community pillar, the issue of armaments has been specified as an area of national government activity and thus exempted from community competence (Article 296, formerly Article 223 in the Treaty of Rome, see also next chapter). The main reason for this exemption of armaments is national foreign and security interests. "Governments became aware that, if they had a single market in armaments in the EC/EU, they would also need a common arms export policy—to prevent EU governments with more relaxed arms export policies from buying equipment from other EU states with tighter controls and then re-exporting with an appropriate price markup. Also, with a single market but without a common arms export policy, it would be likely in the long term that defense production would be concentrated by European defense firms in countries with the loosest policies for export beyond the Union" (Taylor and Schmidt 1997, 122-23).

A European armaments market would therefore open many political and security-sensitive questions. However, in 1996 the Commission presented a communication in which it raised both market and defense aspects of the European defense industry and the issue of how the EU should cooperate on armaments (COM (96) 10). Although the defense frame is toned down in the document, it is quite clear that the COM document on the European defense industry was influenced not only by the Directorate General (DG) for Industrial Affairs (III) (chapter 4) but also by the DG for External Political Relations (IA).[5] It is stated that the intergovernmental conference, which began at the end of March 1996, would discuss developments "concerning common security and defense policies, including the armaments aspects. . . . the Conference should consider how to encourage the development of European operational capabilities, how to promote closer European cooperation in the field of armaments and how to ensure greater coherence of action in the military field with the political, economic or humanitarian aspects of European crisis management" (COM (96) 10 Final: 12). Indeed, according to interviews, the document was the result of teamwork by the two general directorates, which caused tensions between two different approaches on the restructuring of the defense industry (Interviews with senior officials within DG IA and DG III 1997, see also chapter 4). DG IA, which deals with CFSP matters, did not pursue a communitarization (supranational) approach of the defense sector. In its view, the question of the restructuring of the defense industry is a CFSP issue, that is, it belongs to the second pillar in which cooperation based on intergovernmentalism is the rule. An early version of the report was already presented in 1993, within the Commission. This early draft, in the view of DG IA, placed too much emphasis on the economic aspects of the restructuration of the defense industry (Interview, senior official in DG IA February 1997). In early 1995 CFSP aspects were incorporated and the final report was more influenced by DG IA's line of policies.[6] Although the communication is not very strong when it comes to defense and security issues, it is obvious that there have been two competing frames within the Commission on the issue of armaments (Mörth 2000a; see also chapters 4 and 5).

The security aspects of the issue of armaments were presented to DG IA in a research report in March 1997—"The Role of the Armaments Industry in Supporting the Preparation and Conduct of Military Operations" (Taylor and Schmidt 1997). Although the study is not the official view of DG IA, it gives a picture of the kinds of problems that are connected to a CFSP perspective. The central question in the study is: To what extent are nations becoming dependent on defense industrial support, and in what ways will this dependence affect the ability to use military force? "Will the industry-led drive for more cost efficiency by transnational specialization and worksharing in Europe contradict the current state of affairs in European defense policy by undermining the capability of individual European nations to act militarily without active support by other European countries?" (Taylor and Schmidt 1997, 4). The starting point for the study for DG IA is that military operations need considerable industrial support. The experience of the Gulf War and defense budget trends since then show that military dependence on timely industrial support will persist or even increase. The budget reductions in Western countries will prevent "Western armed forces being comprehensively modernized with equipment of very high reliability" (Taylor and Schmidt 1997, 43). Thus, to engage in military operations there is a need for a modernization of national armaments. This modernization is costly, which means that nations must cooperate. Multinational operations also require common equipment or standardized subsystems, components, munitions, etc. The authors of the study take the view that the armaments policy in Europe is inefficient and that there is a "real threat to the survival of a European defense industrial base that is able to provide European forces with state-of-the-art equipment" (Taylor et al. 1997: 8). "Under these circumstances the big challenge for European defense industrial policy is to avoid a situation where there are only two policy options left: either to buy less capable but national or European equipment with security of supply assured but military superiority undermined, or to buy highly capable US equipment providing military superiority over potential opponents but with security of supply being dependent on US willingness and capability to support European forces in specific scenarios. If decision-makers want to avoid this choice and thus improve the prospects of a globally competitive European DTIB,[7] European Arms and defense industrial co-operation has to be improved dramatically and quickly" (Taylor and Schmidt 1997, 48-49). The report also notes the tension that exists between the rules of an economic market (which means that stocks are costly) and the national security need for a reliable supplier of timely defense support in a military operation. Clearly, the report predicts the important problems that have been discussed since the Amsterdam Treaty in connection with the issue of armaments within a defense frame.

The Council of Ministers never considered the communication from the Commission. The political room for maneuver on defense issues was very limited at the time, and the EU lacked an institutional and organizational capacity for dealing with issues related to defense. This situation created a political problem when the Commission brought the issue of armaments onto the EU's political agenda. As one of my interviewees put it, "When the Commission presented its communication on the restructuration of the defense industry in January 1996 the issue of the European defense industry was formally a matter for the foreign ministers, but they did not find the time to discuss it and the defense ministers were not allowed to meet" (Interview, senior official in DG III September 1999). However, in July 1995 a working group in the Council on the issue of armaments was established—POLARM (the ad hoc Working Party on a European

Armaments Policy).[8] The tasks of the group were studying "the options for armaments policy, making recommendations for further action and proposing specific measures within the EU's jurisdiction" (CEPS 1999, 17). In December 1996 it was stated by the working group that armaments is a very special sector for European cooperation since the defense industry is a national strategic industry (Coreper on 10 December 1996, in COM (97) 583). One of its characterizations, according to POLARM, is that the governments are the only customers, and in some cases major owners of defense industries as well. "Also, relations between the government and defense-related companies differ considerable between Member States, for example the extent of State ownership, the extent of funding R & D, etc. The role of governments explains many of the characteristics of this sector and has a major influence on the way it is built up and restructured. At the same time, the industry itself has a specific role and responsibilities in this regard" (Coreper on 10 December 1996, in COM (97) 583: 6).

In May 1997 the so-called Titely report was voted in plenary session in the European Parliament as a reaction to the communication from the Commission (European Parliament 1997). The report was discussed not only in the Committee on Foreign Affairs, Security, and Defense Policy, but also in the Committee on Research, Technological Development and Energy.[9] In the resolution it is stated that Article 296 "should not be deleted until a common foreign and security policy including a restrictive arms export policy has been established" (European Parliament 1997, paragraph 17*)*.[10] Hence, the political decisions on these matters must precede a deregulation of the defense sector. Already in the so-called Poettering report in 1990, the Parliament discussed the abolition of Article 296 and the creation of an independent European agency on armaments (Mezzadri 2000; see also chapter 5).[11]

From this reading it is clear that a defense frame on the issue of armaments has not been discussed by various organizations within the EU until recently. The move of general political process in the late 1990s and in the early part of this century toward a more coherent and extended EU foreign and security policy has, however, pushed the issue of European armaments forward. Already in the Treaty on European Union it is stated that the "CFSP shall include all questions related to the security of the Union." In the WEU declaration, which is annexed to the Maastricht Treaty, a closer relationship between the EU and the WEU on the issue of armaments is mentioned—"enhanced cooperation in the field of armaments with the aim of creating a European armaments agency" (TEU Annex 1, 4; on the question of a European Armaments Agency see below and chapter 5).

The Amsterdam Treaty supports a defense frame on the issue of armaments. Article 17 (formerly J.7) of the Amsterdam Treaty declares that the "progressive framing of a common defense policy will be supported, as Member States consider appropriate, by cooperation between them in the field of armaments." Thus, an intergovernmental way (that is, within the second pillar) of handling the issue of armaments is emphasized. This sole reference in the Amsterdam Treaty to the issue of armaments has, however, been interpreted as rather weak and inadequate (CEPS 1999). Two years later, in the Cologne Declaration, the issue is given more political attention. It is recognized that conflict prevention and crisis management in the European Union require appropriate capabilities and instruments. The efforts to bring together multinational European forces require that the EU "undertake sustained efforts to strengthen the industrial and technological defense base, which we want to be competitive and dynamic. We are determined to foster the restructuring of the European defense industries amongst those States involved. With industry we will therefore work towards closer and more efficient defense industry collaboration. We will seek further progress in the harmonization of military requirements and the planning and procurements of arms, as Member States consider appropriate" (Annex III, paragraph 2). In the Nice Treaty it is stated that "the progressive framing of a common defense policy will be supported, as Member States consider appropriate, by cooperation between them in the field of armaments" (TEU Article 17).

This interlinkage between a European capacity on defense and the defense industry is also underlined by the national EU governments. British Defense Minister Geoff Hoon has declared, "The main focus of our work should be the development of the military capabilities of European nations. Clearly, developments in Europe's security and defense policy will amount to very little without better military capabilities to underpin our foreign policy objectives. . . . We must have deployability, sustainability, flexibility, mobility, survivability and interoperability" (Hoon 2000).[12] In a joint statement, the French and German governments declared that "France and Germany intend to achieve further progress in the area of European armaments policy, an integral part of the European security and defense policy. In this respect, they welcome the conclusion of the Letter of Intent (LOI) negotiations and the forthcoming signing of the framework agreement" (Franco-German Defence and Security Council, 2000). Hence, the new military tasks and threats that are identified will affect the structure of the defense industry and of the products manufactured by that industry. "While the cold war emphasized building up stockpiles of weapons, conflicts in the Gulf and Kosovo have stressed the importance of responding quickly, picking out targets, and ensuring effective communications" (Peter Thal Larsen 1999)

The issue of armaments is not explicitly mentioned in the Helsinki European Council meeting in December 1999. The summit and the Presidency Conclusions, however, draw on the political commitments made at Cologne in June 1999. The decision to develop European capabilities in the form of 60,000 troops able to function within 60 days will entail technical and organizational standardization. It is stated that "these forces should be militarily self-sustaining with the necessary command, control and intelligence capabilities, logistics, other combat support services and additionally, as appropriate, air and naval elements" (Helsinki European Council 1999, Presidency Conclusions).

So, what we can see today is that the European Union, including the Commission, the Council of Ministers, and the Parliament, are discussing the issue of armaments and how this issue is an integral part of the emerging European defense policy. Initially, the political controversies seemed to be most salient in the Commission. The Council and the heads of governments, on the other hand, had a clear view of where the issue belongs, namely, to the defense policy process and to the intergovernmental pillar of the European Union. This does not mean, however, that there are no political tensions within a defense organizational field over how to cooperate on armaments. The tensions have not been very salient in the EU but have taken place outside the Union (see below and chapter 5).

NATO

NATO is in a new security and defense situation and is struggling to find credible approaches to unfamiliar security problems, for instance, in the former Yugoslavia. Thus, NATO is no longer the classic military organization it once was, based on territorial integrity and security guarantees. The alliance's new role—a cooperative security role—includes cooperation and dialogue on a multitude of topics to "reduce mistrust, deepen understanding of the preoccupations of others, provide for reliable and continuous communication, and promote a sense of shared responsibility for security" (Yost 1998, 270). The new role for NATO also involves activities to prepare some countries for membership, "cultivating capabilities for joint action in crisis management and peace operations, and promoting economic reform and democratization, including civilian control of the military" (Yost 1998, 270). Indeed, multifunctional peacekeeping requires a more civilian-oriented organization and thus puts new demands on NATO on top of the purely military ones.

NATO is involved in an increasing number of collaborative military projects. Many of these are organized under the Combined Joint Task Force (CJTF), which allows for "ad hoc combinations of NATO and WEU forces for specific missions under a single command" (Yost 1998, 270). The CJTFs are thus central to NATO's new role in European security policy. These new forces, which are multinational, aim to upgrade NATO's ability to conduct non–Article 5 operations, that is, operations that do not involve the collective defense of the territories of NATO states. The distinction between Article 5 and non–Article 5 operations seems rather unclear, however. In June 1996 it was agreed that CJTFs could be made available to the WEU. "Consequently, the idea of a European Security and Defense Identity (ESDI) within NATO has become closely linked with the development of the CJTFs" (Sjursen 1998, 102). Once again, political authority was unclear between NATO and the WEU. The phrase "separable but not separate" is a tool to get around this problem of authority and possible duplication between the two organizations (Yost 1998).

CJTFs will give the allies a standby capability for peacekeeping and peace enforcement. This new type of initiative gives NATO the new tasks of crisis management. Crisis management refers to military operations that can have civilian components. A popular notion is that crisis management requires smaller and more flexible multinational forces to respond to contingencies over a geographical area. NATO must thus adapt to these new missions. The CJTF will operate under agreed NATO standard operating procedures (SOPs) and standardization agreements (STANAGs). Numerous STANAGs have been refined over forty years of collective defense operations. Furthermore, non–NATO nations engaging in CJTF operations must also be proficient in these procedures in order to participate successfully in various contingencies. The partnership countries to some extent participate in the work of different STANAGs committees. The work within Partnership for Peace (PFP) plays a crucial role "by developing the capability for non–NATO states to integrate smoothly into CJTFs" (Yost 1998, 210). This means that the cooperation within the Partnership for Peace framework requires standardization of working methods along with standardization of armaments. It means that NATO/ PFP is working on a new operational concept. This new operational concept will also include participation in the Planning and Review Process (PARP), that is, defense planning.

NATO's definition of interoperability has traditionally emphasized the effective operation of technical systems. The requirements of interoperability have changed since the launch of Partnership for Peace, however. It now also includes the "training of personnel and units in NATO doctrine, procedures and practices which are capable of working effectively within NATO or NATO-led organizations on specific operations" (Naumann 1996, 17). One important rationale behind PFP is a desire to develop cooperative military relations with NATO "for the purpose of planning, training and exercising in order to strengthen their ability to undertake missions in such fields as peacekeeping, search and rescue and humanitarian operations" (Naumann 1996, 19). The question of interoperability is not new in NATO's history. The organization has struggled with the arms standardization issue for almost fifty years. NATO has also had the vision of creating an arms market (Taft and Taylor 1992, Hayward 1997). In 1966 the Conference of National Armaments Directors (CNAD) was established. The CNAD is responsible for cooperation among NATO countries on armaments matters, and its principal activities include "defining the conditions for standardization and interoperability that equipment belonging to the armed forces of the member countries of the Atlantic Alliance must fulfill to achieve effective joint intervention" (Ministère de la Défense 1999, 6). The directors meet twice a year. The day-to-day work is handled by the National Armaments Directors (NAD) representatives. The CNAD strives to coordinate the political, economic, and technical aspects of NATO forces armaments. The partnership countries participate to some extent in this work and in the CNAD's various committees. So, the issues of cooperation on armaments are dealt with primarily by the CNAD "and [are] based on an information exchange process, which seeks agreement between nations and the major NATO commanders (MNCs) on harmonized operational requirements to promote cooperative programs" (CEPS 1999, 20). Industry is involved in the CNAD through the NATO Industrial Advisory Group (NIAG), which to a large extent coincides with the grouping of EDIG and the European Association of Aerospace Manufacturers (AECMA) (see chapter 4.5).

In 1981 a Phased Armaments Programming System (PAPS) was introduced. PAPS is a review system designed to identify areas of co-operation.

"It is based on Equipment Replacement Schedules provided annually by participating nations" (CEPS 1999, 20). These schedules are examined by the CNAD and others within NATO (such as the MNCs). In the mid-1980s CNAD produced an Armaments Cooperation Improvement Strategy (NAC 1997). "In its recommendations to the NAC (North Atlantic Council), it called for the establishment of a process of armaments planning to guide Alliance armaments co-operation efforts. Thus far, NATO armaments planning had been both cumbersome and separate from the main NATO planning system" (CEPS 1999, 21). NATO's work on standardization and interoperability issues concerning armaments took an interesting turn in 1996 when CNAD approved a proposal by the Conference's Permanent Chairman to initiate a wide-ranging "NATO Armaments Review with a focus on the CAP" (NAC 1997). CAPS stands for the conventional armaments planning system, which was accepted in 1989 by NAC to promote defense procurement harmonization and is now generally accepted within NATO (CEPS 1999). The Armaments Group has presented several reports, in 1997 and 1998 (17 December 1997, 16 April 1998, and 21 October 1998). These reports were submitted to the CNAD for approval and were forwarded to the North Atlantic Council in March 2000.

The reports are rather complex in substance, but the overall aim seems to be to form a "NATO Armaments Community" (NAC 1998a). The rationale behind the decision to launch a major study "was the pressing need to take a view on the longer term implications of the new security environment for the Alliance in the armaments field and determine what roles and functions NATO should play in this area in the future, while, at the same time, fulfilling the remit issued by the North Atlantic Council in 1993 to assess the performance of the Conventional Armaments Planning System (CAPS) upon completion of the 1996-1997 CAPS cycle" (NAC 1997, 1-2). The general principles and guidelines concerning the alliance's armaments-related activities are only briefly touched on in a couple of its charter documents. Hence, what is needed is the formation of a general armaments framework within NATO that will coordinate the various internal activities.

Two major key processes are perceived as central in the reports: (1) the harmonization of military requirements (to fulfill the alliance tasks), which entails a collective defense planning process; and (2) materiel standardization to enhance overall interoperability among alliance forces and between the alliance and its partners (NAC 1997). The need for harmonization of military requirements is emphasized in the

1997 report due to the new defense responsibilities that are entrusted to the alliance by its members. It is stated that NATO has an important role in facilitating "national or NATO acquisition, as appropriate and without duplication of efforts elsewhere, of those armaments-related military capabilities without which the different national contributions to Alliance forces cannot operate, train or exercise together effectively" (NAC 1997, 5). This is because the security environment has changed "which led to reduced defense budgets in most member states; a new Alliance Strategic Concept and hence new missions for NATO; and significant changes to Alliance force structures. One of the most significant changes to those force structures has been a greatly increased reliance on multinational forces" (NAC 1998a, Annex C, 1).

The harmonization of military requirements is a key factor in enhancing coordinated acquisition (NAC 1998a). By that time, there were two NATO agencies with "the administrative and legal ability to place general purpose contracts on behalf of nations, the NC3A and NAMSA" (NAC 1998a, 4).[13] A more general framework and organizational set-up was suggested—the "NATO Materiel Acquisition and Support Program" (NAMASP)—in which a new committee would be created—the "NATO Materiel Acquisition and Support Committee"—NAMASC (NAC 1998a, 4). This new committee will coordinate acquisition, based on national plans, and enhance the dialogue between military leaders, policy officials, and industry (NAC 1998a, B6). Thus, acquisitions should be based primarily on national plans in order to achieve interoperability. This type of acquisition is called coordinated acquisition, defined as "any form of action initiated by governments and/or their armed services to co-operate at any point or points in the acquisition process. Coordinating national acquisition focuses on those functions of acquisition management which can be performed in common more efficiently and cost effectively than can be achieved by countries acting individually" (NAC, 1998a Annex B, 4). Common acquisition is also mentioned, but such measures will respect national sovereignty on these issues (NAC 1998a). It is, however, stated that coordinated acquisitions can be facilitated by some form of common "structures, procedures and processes"(NAC April 1998, Annex B). "More cost-effective defence acquisition will require structures, procedures and processes which can facilitate coordination" (NAC 1998a). These measures will, however, "respect national sovereignty in defence procurement" (NAC 1998a).

There are several organizations within NATO that deal with issues of standardization and harmonization, for instance, the Military Agency for Standardization (MAS) and the NATO Standardization Organization (NSO) which was established in 1995.[14] The reports therefore discuss how this internal activity on armaments issues can be more coordinated to create a NATO armaments community. "There is an ever growing market in the world for military equipment which can claim it is 'NATO standard'" (NAC, 1998a, C8). The creation of a "NATO standard" is compared with the extensive standardization work within the EU. "The EU standard, which must be met to receive a certificate of conformity, is analogous to our proposed NATO standard and the detailed civilian standards are our STANAGs. The process whereby it is confirmed that the military equipment conforms to the NATO standard is one possible way of verifying implementation of STANAGs and certifying that the equipment is interoperable with other related NATO equipment" (NAC 1998a, Annex C, 8). NATO has for almost forty years been attempting to establish agreed materiel standards. Hundreds of STANAGs have been promulgated. "Many shortcomings continue to exist, primarily the inability of the STANAG development and ratification processes to keep up with the pace of technological change in key areas such as communications, data transmission logistics, etcetera" (NAC 1997, 12).

Thus, the activity within NATO on the issue of armaments is extensive. This work goes back to the 1950s, but the end of the Cold War and the new military tasks have revitalized and deepened NATO's discussion of a NATO armaments community. It is, however, difficult to predict the outcome of this complex process since the issue of armaments is dependent on the overall political process on the military role of NATO.

To sum up, it is striking that, despite all these activities, so little coordination has in practice been achieved in this area. There are, to be sure, multiple explanations for this inertia, but the fact that NATO remains a fairly weak organization with an intergovernmental decision-making structure, combined with the remaining national sensitivity involved, provides at least part of the explanation. These factors alone, however, cannot explain the problems of cooperation within NATO on the issue of armaments. One additional reason seems to be that NATO remains an enclave, a strong alliance to be sure, but not part of any ongoing political and economic integration process. It is in this latter respect that the difference between NATO and the European process is especially instructive. This is a very important point, that is, how the market- and defense-related processes seem to be mutually constitutive, and I come back to it in chapter 5.

The WEU: WEAG/WEAO

The European states have for many years pursued a policy of opening national armaments markets to cross–border competition and to promote standardization in equipment. In 1968 an informal group of NATO European defense ministers was set up to deal with armaments cooperation at the European level (CEPS 1999). This group was replaced in 1976 when the IEPG was created, which brought together the European defense ministers and their National Armaments Directors. IEPG, for instance, has promoted collaboration in defense research with European Cooperation for the Long Term in Defense (EUCLID), which was launched in 1989 (Bauer 1992).

In 1984 the organization of the WEU was strengthened and it was given a clear mandate to deal with the development of armaments. The so–called Standing Armaments Committee (SAC) was to develop activities within this area in cooperation with NATO's CNAD (see above) and the IEPG. The issue of armaments was too sensitive to be incorporated into the EU, and there were fears of encouraging supranationalism within this sensitive area (Bauer 1992). The WEU's intergovernmental character "reassured states that it would not undermine their sovereignty and made it more acceptable than the EC" (Bauer 1992, 26). The WEU was also separated from the EU and became the EU's consultative council in defense industrial matters. In 1990 the European Defense Industrial Group (EDIG) was established, which was closely linked to the work within IEPG (CEPS 1996). The representatives of EDIG are also part of NIAG in NATO (see above). Two years later, in 1992, the European defense ministers decided to transfer the IEPG's function to the WEU in the Western European Armaments Group. Indeed, the establishment of WEAG "marked WEU's first serious involvement in the armaments sector" (CEPS 1999, 18). The formal link between the WEU and WEAG is, however, weak. WEAG is not linked to the WEU's charter and lacks a legal personality. WEAG is sometimes referred to as a forum, and its activities are based on four basic cooperation principles:

- All 13 nations should be entitled to participate fully and with the same rights and responsibilities, in any European armaments cooperation forum.

- There should be a single European armaments cooperation forum.
- Armaments cooperation in Europe should be managed by the National Armaments Directors of all the 13 nations, who will be accountable to the Ministers of Defense of those governments.
- The existing links with NATO and EDIG should be maintained.[15]

An important principle in WEAG is that decisions are taken unanimously by the participating states.[16] Another essential principle in the cooperation is the principle of *juste retour*, that is, for every program, each state shall get returns in proportion to the money it has invested.

The objectives that were agreed upon in 1992 were the following: "more efficient use of resources through, inter alia, increased harmonization of requirements; the opening up of national defense markets to cross-border competition; to strengthen the European defense technological and industrial base; cooperation in research and development."[17]

WEAG falls under the responsibility of the ministers of defense, and they meet once a year. The chairmanship rotates between member nations and is normally held for two years (WEU, Secretariat-General). The executive work is run by the NADs, who meet twice a year. Every country also has a Permanent Representative of the NADs in Brussels. So, the WEAG staff group consists of the representatives of the thirteen National Armaments Directors. The Armaments Secretariat within the WEU Secretariat–General works under the authority of WEAG National Armaments Directors and assists the chair nation. Thus the secretariat is dependent on a mandate from the NADs and the ministers (Interview with Swedish NAD Krister Andrén 1998). WEAG is divided into three panels—Panel I: Equipment Programs; Panel II: Research and Technology; and Panel III: Procedures and Economic Matters. These panels report to semiannual meetings of the NADs, who, in turn, report to the defense ministers.

In 1995 the NADs agreed to create a new organization the Western European Armaments Organization. It was formally created by the WEU Ministerial Council on the basis of a proposal by the ministers for defense of the WEAG countries, meeting at Ostend on 19 November 1996. Those thirteen countries participate on an equal footing in WEAO activities. The WEAO is the first European armaments body with international legal personality. Its initial task is the management of the research and technology activities carried out under WEAG. The responsibility for managing EUCLID has been transferred to WEAO (Hayward 1997). In 1996 the WEAG ministers signed the Memorandum of Understanding, called

Technology Arrangement for Laboratories for Defense European Science (THALES), "which provides for improved mechanisms for implementing government-funded research programs and information exchanges" (CEPS 1999, 20).

The Board of Directors of the Western European Armaments Organization, consisting of WEAG NADs or their delegated representatives, held its inaugural meeting on 7 March 1997 (WEU Secretariat-General). The Research Cell, which had already been established in the WEU, became the initial executive body of WEAO in April 1997 and consequently has the capacity to place contracts. WEAO is an executive organ of WEAG and a subsidiary body of the WEU and thus shares the international legal personality of WEU (in contrast to WEAG). However, the legal and political relationships between WEAG and WEAO are complex, especially since the decision was made to incorporate the WEU into the EU. An important question is whether this means that WEAG and WEAO will be abolished as well (chapter 5).

Since both organizations, especially WEAO, are very loose and fluid and can be regarded as paper constructions, it is difficult to get a clear picture of these two bodies. They can be regarded as frameworks for cooperation that can develop both in depth and breadth. The WEAG/WEAO framework has been discussed as the beginning of a European armaments agency (CEPS 1999). "When WEAG Defence Ministers decide that the conditions have been met to proceed to a full European armaments agency (in accordance with Article J.4 of the TEU), the Research Cell will be absorbed into the executive body of the EAA" (CEPS 1999, 20). This establishment of a European Armaments Agency (EAA) is politically sanctioned in the WEU declaration of 10 December 1991, which was annexed to the Maastricht Treaty. An ad hoc study group was created in 1993 and the objective was to "examine all matters related to the possible creation of a European Armaments Agency."[18] This work led to the agreement by the ministers to establish the Western European Armaments Organization. In November 1997, the WEAG ministers agreed to develop a plan, including a timetable, for creating a European Armaments Agency. The plan was written by the chairmanship of WEAG with the assistance of a group of national experts. In a 1998 masterplan report on a European Armaments Agency, it is noted that the plan should be seen in the light of the Commission's support of common harmonized rules and regulations for the armaments sector. This plan was politically sanctioned in a meeting in Rome in November 1998 by the WEAG ministers.

The matter of the competencies and functions of the planned EAA is a politically controversial issue. In the masterplan report it is stated that substantial concrete tasks should be "increasingly delegated from Nations to the EAA" (Masterplan 1998, 2). The agency will be independent in its decisions for the tasks as delegated by the governments and subject to control by WEAO or by the national participants in various programs (Masterplan 1998). The procurement activities by the agency will be based on WEAG objectives, principles, etc., and achieve the "aims of the WEAO as stated in Paragraph 6 of the charter, namely to assist in promoting and enhancing European armaments cooperation, strengthening the European defense and technology base and creating a European armaments market" (Masterplan 1998, last page). The masterplan report seems to support the creation of an independent agency subordinated to WEAG/WEAO. A new proposal was, however, presented in mid–October 1999 in which the main author of the report, General Andries Schlieper, chairman of the WEAG, argued for an umbrella structure and stated that various organizations were compatible with each other (Schlieper, interview, 1999; see also chapter 5). At their meeting in Luxembourg on 22 November 1999 the defense ministers "endorsed the way ahead to continue work towards the implementation of the European Armaments Agency in accordance with the Masterplan" (Pierre Delhotte, letter, 30 November 1999). A decision on whether to implement the would-be agency will be taken at the end of 2001 (Schmitt 2000b). The ambition is to establish an EAA in 2002. Its central function will be procurement (Executive Summary of the Activities of the GNE on the Masterplan for an EAA, November 1999). The range of the activities by the agency could also include research and technology, "studies, management of assets and other activities like in-service support and the formulation of design specifications" (WEAG, Executive Summary of the Activities of the GNE on the Masterplan for an EAA, November 1999).

To sum up, the European activities outside the EU framework on the issue of armaments had already begun at the end of the 1960s. The revitalization of the WEU in 1984 and onward entailed making European cooperation on armaments part of the WEU's new tasks. This means that the European defense ministers have been able to discuss the issue of armaments within the framework of the WEU. Two new bodies have been created with the aim of dealing with the issue of armaments, namely, WEAG and WEAO. They have been considered as the basis for a would—be European Agency on Armaments, especially concerning the procurement of armaments. The relationships between the WEU, WEAG, and WEAO are complex. This is especially the case between the WEU and WEAG since WEAG is not based on the WEU charter and has a rather vague and weak legal status.

Concluding Analysis

In various intergovernmental documents it is stated that the end of the Cold War requires multinational and multifunctional forces. The new types of military operations make interoperability and standardization necessary. Thus, a closer European cooperation on armaments is important. Historically, the EU, the WEU, and NATO—the so-called security triangle—have had complex relationships with each other. The political process toward a European armaments policy has thus been dependent on the general CFSP process and the political break-through that came with the Amsterdam Treaty and the Cologne process. Indeed, the political context has dramatically changed during recent years and has activated a defense frame on the issue of armaments. The treaties and texts from the European Council summits legitimize activities that put the issue of armaments into a defense frame. Several of my interviewees talked about the "Cologne spirit" that created room for maneuver in a policy area that has traditionally been politically closed. The war in Kosovo, the declarations in Saint Malo, Cologne, etc., have clearly opened a political window for pursuing a defense frame on the issue of armaments in the European Union.

The defense frame on armaments is part of the general question of Europe's emerging defense policy. It has therefore primarily activated the intergovernmental part of the EU: the European Council, the Council of Ministers, NATO, and the WEU. Parts of the European Commission have also been active in framing the issue of armaments as belonging to the defense field. The regulative rules of the field concern the second pillar of the Union; the constitutive rules are the notion of the need for a European defense policy and that this policy is to be decided in an intergovernmental way. Consequently, the diagnosis of Europe's problem in a defense field focuses on military capacity problems and the need for a common defense policy. An emphasis on the intergovernmental pillar of the European Union is also ubiquitous, as is an emphasis on the need to secure national sovereignty.

Chapter 4 shows that the issue of armaments has also been part of a different organizational field. I would even argue that the formation of European cooperation on armaments in the late 1990s and early part of this century started in a field other than the defense field. This field is called the market field since it consists of organizations that deal with issues within the first pillar of the European Union. We now turn to an analysis of this organizational field and see how the issue of armaments started its journey in the European process of organizing in the early and mid-1990s.

Notes

1. The countries behind the report were France, West Germany, Belgium, Italy, Luxembourg, and the Netherlands (Sloan 1985).

2. The plan was initiated by the German foreign minister, Genscher, and the Italian foreign minister, Colombo (Sloan 1985).

3. The "Solemn Declaration on European Union" in 1983 states: "To strengthen and develop European Political Cooperation through the elaboration and adoption of joint positions and joint action, on the basis of intensified consultations, in the area of foreign policy, including the coordination of the positions of Member-States on the political and economic aspects of security, so as to promote and facilitate the progressive development of such positions and actions on a growing number of foreign policy fields" (Sloan 1985,181).

4. See also other summits: the Franco–German summit in Potsdam in December 1998, Italian-British summit in July 1999, and the Franco-Italian summit in September 1999 (Mezzadri 2000).

5. During the autumn of 1999 the General Directorates were renamed and partly reorganized. DG IA— external affairs is now External Relations DG.

6. This information is partly contested within DG III, which claims that DG IA's role in the final phase was rather limited.

7. DTIB—Defence Technology Industrial Base.

8. In 1995 a working group on export issues—COARM—was established.

9. It has also been discussed in the Committee on Economic and Monetary Affairs and Industrial Policy and the Committee on External Economic Relations and the Committee on Institutional Affairs.

10. The ambition to arrogate to itself jurisdiction in armaments collaboration goes back to the Klepsch report on the subject to the European Parliament many years ago (CEPS 1997; Mezzadri 2000).

11. In 1997 the Commission issued another communication on armaments.

Since this communication is based on a different logic, or a reframing of the issue, this communication and the political responses to it are analyzed in chapter 5.

12. The Letter of Intent (LOI) is analyzed in chapter 5.

13. NAMSA is focused on logistics.

14. NSO consists of three bodies: The NATO Committee for Standardization (NCS) which is the senior authority on general NATO standardization matters; The NATO Standardization Liaison Board (NSLB) and The Office of NATO Standardization (Lindgren 1998, see also Ferrari 1995).

15. www.weu.int/weag/eng/info/weag,htm, 16 November 1997.

16. The various membership statuses in WEAG have been rather confusing for me as a researcher and for the states involved. WEAG was recently expanded from thirteen to nineteen states. In November 2000 Sweden, Finland, Poland, the Czech Republic, Hungary, and Austria were elected into WEAG. The term membership is not completely accurate, since several of my interviewees have argued that it is impossible to be a member of WEAG due to the fact that WEAG is not a real organization (!). It is more common, instead, to talk about participating states in WEAG.

17. www.weu.int/weag/eng/info/weag,htm, 16 November 1997.

18. www.weu.int/weag/eng/info/weag.htm.

Chapter 4
The Market Field

With reference to the defense industry, the U.S. led the way in defense rationalizations while we were slow starters.

—Lars Josefsson, Senior Executive
Vice President, Saab

This chapter focuses on how the issue of armaments has been framed in an organizational field in which Europe's civilian technological and industrial competitiveness is highlighted. I argue that even though the political breakthrough for the issue of armaments occurred in the wake of the end of the Cold War and the Cologne process, the issue has also long been considered to be a question of Europe's (civilian) economic and technological capacity in relation to the United States. This is especially salient in the 1990s and in the early years of the new millenium, when the need for the EU to address the issue of European economic competitiveness and technological development is more articulated than in any earlier period of the EU's history. The political emphasis on competitiveness can be illustrated by the European Council's meeting in Lisbon in March 2000 in which it was stated, "The European Union is confronted with a quantum shift resulting from globalisation and the challenges of a new knowledge-driven economy. These changes are affecting every aspect of people's lives and require a radical transformation of the European economy" (European Council 2000, Presidency Conclusions, paragraph 1). For that reason, the Council outlined a "new strategic goal for the next decade: to become the most competitive and dynamic knowledge-based economy in the world capable of sustainable economic growth with more and better jobs and greater social cohesion" (European Council 2000, Presidency Conclusions, paragraph 5).

This chapter shows that the perception of the need to strengthen Europe's economic and technological competitiveness is a regular feature in EU politics. Since the creation of the European Communities in the 1950s, the issue of Europe's technological and economic competitiveness in relation to the United States has been on the political agenda. Three waves of technology-gap fever can be identified (Sandholtz 1992). The first occurred in the mid-1960s, the second in the early 1980s, and the third during the 1990s. The next section of the chapter discusses how these three waves of a perceived technology gap have been presented and handled by the EU (the Council of Ministers, the European Council, the European Commission, and the European Parliament). The second section analyses how the issue of armaments has been part of this technology-gap fever in the 1990s and how a market frame activates the first pillar of the European Union and thus raises a different set of questions and processes than is the case with the defense frame within the defense field.

Europe's Competitiveness—Three Waves of Technology-Gap Fever[1]

The First Wave

Research and technological development (RTD) has been part of the EU since the 1950s and the Euratom Treaty. The Treaty of the Coal and Steel Community had already referred to the importance of technological and economic research, and in 1955 the first RTD program was launched (Mörth 1996). This cooperation was not based on any article in the treaties but on the general Article 235 in the Treaty of Rome.[2] This article provided "a range of policy powers which could be used to determine the regulatory framework and market conditions for European industry. Thus, competition policy, freedom of capital and labor movements, the right of establishment, customs union, harmonization of national laws, and state aids fell within the treaty's competence. But they were not subsumed under a general framework for industrial policy" (Sharp and Shearman 1987, 26).

In the mid-1960s technology policy and industrial policy began to be linked, both in national politics and at the EU level. "At the same time, a panicky debate erupted in Europe over technology gaps that left European industries dangerously behind the American competitors" (Sandholtz 1992, 70). National programs were launched, and the aim was to create national high-tech champions (Sandholtz 1992, 70; Mörth 1996). The diagnosis of a technology gap was presented by national actors such as governments, and by European and international organizations, especially the Organization for Economic Cooperation and Development (OECD) (Sandholtz 1992; Mörth 1996). The well-known book by Jean-Jacques Servan–Schreiber, *The American Challenge,* illustrates the general sense of a European malaise towards the United States that existed in the late 1960s (Servan–Schreiber 1968). "La puissance américaine n'est plus celle que nous avons connu après la guerre. Elle a changé d'échelle et, pour ainsi dire, de nature. Nous la découvrons parce qu'elle débarque ici. Tant mieux. Le choc est rude" (Servan–Schreiber 1968, 39). From this diagnostic framing, the author continues with suggestions for European strategies. "Formation de grandes unités industrielles capables, non seulement par leur taille, mais *par leur gestion,* de rivaliser avec les géants américains . . . .Choix des 'grandes opérations' de techniques de pointe qui préserveront, *sur l'essentiel,* un avenir autonome pour l'Europe. . . . Un minimum de pouvoir féderal qui puisse être *le promoteur et le garant* des entreprises communautaires" (Servan-Schreiber 1968, 171, emphasis in original). There are many ways in which one could comment on this quotation. Suffice it here to say that the lagging-behind theme has a timeless quality.

Another policy response to the perceived American threat was the decision to create a more coherent RTD and industrial policy at the European level. An important component in this endeavor was a plan to create the Framework Program on RTD in 1983 (see below). This process had started already in the late 1960s when the Council of Ministers adopted RTD plans that were elaborated by the Commission.[3] The Commission advocated supranational RTD and industrial policies, especially those proposed by Commissioner Altiero Spinelli (1970-76), whereas the governments pursued a more intergovernmental policy-making style (Nau 1975; Hodges 1983). The Commission and the governments shared the view, however, that the EC should take a greater responsibility for RTD and industrial issues that were regarded as strategic for the future of Europe. Under the leadership of Pierre Aigrain, French Delegate-General for Scientific and Technical Research, "over forty projects for European–owned companies (to avoid encouraging further US penetration) were put forward under the sectoral headings already identified" (Peterson and Sharp 1998, 31). In 1971 an outline plan was decided for a program of Cooperation in Science and Technology (COST).

The Second Wave

The second wave of technology-gap fever came in the early 1980s, and it was even more precisely articulated in the period after the decision of the Single European Act in 1986/87. The priority for EU industrial and RTD policies was to increase European competitiveness (Barnes and Barnes 1995). National high-tech champions were no longer considered to be enough, and it was instead important to create European-oriented companies and a more coherent EU industrial and RTD policy. The policy responses this time were more focused at the European level, and one reason for this was the political awareness of Japan. "The impetus behind a legal and political mandate for an EU technology policy in the 1980's was the re-emergence of concern about Europe's lagging technological competitiveness—this time not just *vis-à-vis* America, but also in relation to Japan" (Peterson and Sharp 1998, 68).

Three major European RTD programs were launched between 1982 and 1985. Two programs were sponsored by the EU, ESPRIT—European Strategic Program for Research and Development in Information Technology—and RACE—Research and Development in Advanced Communications Technologies for Europe. The commissioner at the time, Etienne Davignon, initiated the programs together with the directors of the twelve largest European information technology companies (Sandholtz 1992; Peterson 1992). The programs were, together with other high-tech programs, part of the EU's framework program for research and technological development, which was decided by the Council in 1983. The third program, EUREKA—European Research Coordination Agency—was initiated by then-French President Mitterrand in 1985 as a first response to the American Strategic Defense Initiative (Mörth 1996). EUREKA was formally organized and financed outside the framework program, but it "responded to the same fears about the status of high technology that motivated ESPRIT and RACE" (Sandholtz 1992, 5). The framework program, COST, and EUREKA comprised a European Technological Community and provided an umbrella for the promotion of a technological fortress Europe (Wyatt-Walter 1995).

The framework program on RTD was first legally formalized under the Single European Act and reinforced in the Treaty on European Union. Article 163 (formerly Article 130f) of the union treaty sets out some general objectives for research in Europe:

1. The Community shall have the objective of strengthening the scientific and technological bases of Community industry and encouraging it to become more competitive at the international level while promoting all the research activities deemed necessary by virtue of other Chapters of this Treaty.
2. For this purpose the Community shall, throughout the Community, encourage undertakings, including small and medium-sized undertakings, research centers and universities in their research and technological development activities of high quality; it shall support their efforts to co-operate with one another, aiming, notably, at enabling undertakings to exploit the internal market potential to the full, in particular through the opening-up of national public contracts, the definition of common standards and the removal of legal fiscal obstacles to that co-operation.
3. All Community activities under this Treaty in the area of research and technological development, including demonstration projects, shall be decided on and implemented in accordance with the provisions of this Title.

Whereas the Single European Act stated that "the Community shall have the objective of strengthening the scientific and technological bases of Community industry and encouraging it to become more competitive at the international level," the Maastricht Treaty adds, "while promoting all the research activities deemed necessary by virtue of other Chapters of this Treaty" (see also Barnes and Barnes 1995; Middlemaas 1995). Thus, the Treaty on European Union (TEU) meant that research activities should become part of other policy areas—a call for horizontal linkages. This formal change in the treaty gave legitimacy to those research activities that were initiated on the basis of Article 308 (formerly Article 235) of the treaty. TEU also stressed the importance of coordination between the Community and member states and gave the Commission a central role in this coordination process (for a discussion on coordination problems see COM (94) 438, final).

The Third Wave

The EU's RTD policy can be characterized in the 1990s and in the early years of this century by the continuous building up of Europe's structural base of civilian power—which has been called soft power (Nye 1990; Nye and Owens 1996). "Do less, do better" was the motto of the former president of the Commission, Jacques Santer, and one of his first initiatives as president of the Commission in January 1995 was to establish an independent group of advisers who would counsel EU leaders on how to improve European competitiveness (Competitiveness Advisory Group). This group consisted of leading industrialists, trade unionists, politicians, and academics. The Competitiveness Advisory Group has published several influential reports on how to enhance the competitiveness of Europe, for instance by opening closed sectors of the European economy such as telecommunications and electricity (Jacquemin and Pench 1997). The image of a declining Europe is still vivid but has somewhat changed focus. The question of Europe's competitiveness was less focused on the technology gap and more on unemployment "and the EU's failure to match either the fast growth-rates of South–East Asia or the faster employment creation of the U.S." (Peterson and Sharp 1998, 12). Thus, the inability to create jobs in Europe was considered to be an important symptom of the region's "declining competitiveness" (Peterson and Sharp 1998, 12). The most comprehensive EU approach toward a common industrial policy emerged out of the white paper "Growth, Competitiveness, Employment," which was published in December 1993 (COM (93) 700). The report identified a number of issues concerning EU competitiveness and aimed to "lay the foundations for sustainable development of the European economies, thereby enabling them to withstand international competition while creating the millions of jobs that are needed" (COM (93) 700, Preamble). The Commission had already, in 1990, presented the so-called Bangemann report (COM (90) 556), which outlined the major problems that the European industry was facing in an increasingly open and competitive environment. It was argued that Europe's competitive position in relation to Japan and the United States had worsened with regard to employment, shares of export markets, R&D, etc. (COM (90) 556).

A key solution to the problems of Europe's competitiveness issue that was addressed in the 1993 white paper was linked to the importance of a strong knowledge-based economy. "In the wake of the globalization of economies and markets, it is no longer possible to divide industry and geographical areas into clearly identified and relatively independent segments" (COM (93) 700, preamble). This was, according to the Commission, most evident within the fields of telecommunications, information technology, consumer electronics, etc., in which strategic alliances between firms are increasing (COM (93) 700: preamble). The former commissioner for industry, Martin Bangemann, has repeatedly emphasized the need for the European nation–states to act as one entity since Europe is competing globally. "The world is now becoming a global economy, thanks to Information and Communications Technology. In the global economy, wealth will only come from our ability to compete. We are in a race for competitiveness . . . competitiveness is not everything. But without competitiveness, everything is nothing" (Bangemann 1996).

The problems of coordination between various political levels and policy activities have also been at the center of various communications during the 1990s. According to the Commission, this was especially a problem within the RTD area in which national policies were still developed largely without reference to one another (COM (93) 700). The lack of coordination was "particularly marked between military and civil research activities in each Member State" (COM (93) 700, 98). The Commission argued that the lack of coordination between various RTD activities and other policy areas was an important factor behind the greatest weakness of Europe's research base, namely, its "limited capacity to convert scientific breakthroughs and technological achievements into industrial and commercial successes" (COM (93) 700, 98). This European weakness was explicitly contrasted to the case of the United States, which, according to the Commission, has succeeded in transforming research accomplishments into the commercial market (COM (93) 700: 98). The diagnosis of Europe's problem vis-à-vis the United States is also well elaborated in the Commission's 1995 green paper on innovation (European Commission 1995). The paper presents the European paradox—Europe is good at research but not at transforming these skills into a competitive advantage. The reason for this paradox is that the European effort is fragmented. In an interview, the former commissioner for research, Edith Cresson, argues, "I am afraid that we are wasting resources by spreading them too thinly over too many fields. This is why, together with my colleagues Commissioners Bangemann and Kinnock, I introduced the Task Forces. Their aim is to strengthen cooperation and coordination between research and industry, and to target our research efforts more precisely" (Cresson 1996). What was needed was a "genuine European strategy for the promotion of innovation" (European Commission 1995, 5; see also Guzzetti 1995).

Clearly, technology policy has thus been closely linked to other policy areas, such as trade, competition, and economic policy (Cini and McGowan 1998). Indeed, by the late 1990s the EU's technology policy had a far clearer and more widely accepted rationale than ever before. "EU-funded research became more closely linked with and responsive to other technology policy actions, such as the liberalization of the telecommunications sector, as well as other EU policies. A new initiative to revitalize the post-Maastricht EU, taken at the end of Delors' presidency, further cemented the position of technology policy as a secure part of the EU's policy repertoire" (Peterson and Sharp 1998, 114). The changed technology policy was evident in the framework program during the 1990s. The redistributive component of the program changed in the early 1990s to become more targeted toward strengthening Europe's competitiveness (Peterson 1995). In April 1997 the Commission presented a proposal for the fifth framework program (F5P) for research and technological development (1998–2002, COM (97), 142). The overall theme of the proposal was how to cope with globalization on a European level by making European research more effective and giving European cooperation added value. The program covered a very broad area, including such issues as the knowledge-based society, employment, economic globalization, European competitiveness, and foreign policy.

The importance of the information society and Europe's technological and economic competitiveness has been a concurrent theme in the Presidency Conclusions of the European Council during the 1990s and in the early years of this century.[4] There has thus been no disagreement between the Commission and the European Council on how to diagnose Europe's problem. One difference, however, between the Commission's communications and the Presidency Conclusions of the European Council is that, whereas the Commission has underlined the problematic relationship between Europe and the United States, the European Council hardly ever explicitly mentions the European malaise towards the United States. This could be explained by the fact that the Commission is more focused on thinking of Europe as an entity and on the matter of how to transform a fragmented European effort into a more coherent actor in a globalize economy.

To sum up, the three waves of technology-gap fever in the EU all share the general notion of a technologically declining Europe. The responses to this perceived threat to Europe's prosperity have consisted of various RTD programs at the European level. The Commission and the European Council have repeatedly stressed that it is important that EU activities be coordinated and that Europe must be seen as an entity. We now turn from this general discourse on Europe's technological and economic competitiveness to the issue of armaments and how the future of Europe's defense industry has become part of the classic lagging–behind theme.

The Commission and the Market Frame on the Issue of Armaments

In January 1996 the Commission took the unusual step of explicitly discussing the difficulty of separating civilian- and defense-related technology and asserting that this fact had to be considered in various ways in EU policymaking (COM (96) 10). The Commission stated, "It has been estimated that technological areas of potential dual-use account for as much as one third of the overall Community research budget. It is therefore not surprising that a number of companies and research organizations known to be active in the defense sector participate in Community programs and that some member-states are encouraging them to do so. Some of these companies are also being consulted in the framework of the Commission's Task Forces to improve the links between research and industry (for example aeronautics)" (COM (96) 10, 20). An important argument for the perceived need for increased linkages between the civilian-and defense-related spheres was the changed dynamics in the technological and industrial sector. Traditionally, it was the military sphere that gave the civilian sphere the technology—the so-called spin-off effect—but the spin-off effect has more or less been replaced by what has been dubbed the spin-in effect (COM (96) 10: 20). This meant that the defense industry was becoming more dependent on civilian industry and civilian RTD programs. The interlinkages between the two spheres were especially evident in the space and aerospace sector, which was in great need of a coordinated approach

(COM (96) 617).

The communication in January 1996, sometimes referred to as the Bangemann report, after the former Commissioner on Industry, raised the issue of incorporating dual-use technology into the framework program for RTD. The Commission argued that it was necessary to consider how, and to what extent, increased civil-defense synergies could be promoted at the European level with the aim of optimizing the overall use of research and development. However, when the Commission presented a formal proposal for the fifth framework program in April 1997, it made no mention of whether dual-use technology should be part of the program, although this matter had been mentioned in an early internal draft and had been raised by several member governments. For instance, in the French position paper of the spring of 1996, the French government took an explicit stand on this issue and declared that it was "desirable to have closer co-ordination between the 5th FP and a redefined EUCLID[5] programme" (www.cordis.lu/fifth/src/ms-se 1.htm#vie). The suggestion that dual-use considerations be taken into account in the framework program proved to be very controversial and sensitive in the Directorate General for Science, Research, Development (DG XII) and the Cresson cabinet (interview with official in DG XII, February 1997). It was also a sensitive issue among some of the member states, a conclusion that can be drawn from the fact that so few governments mentioned dual–use technology in their position papers. The reason for the reluctance of some members of the Commission and the Council of Ministers to explicitly mention dual–use technology in the framework program seemed to have been that this could have opened "Pandora's box," putting into question the civilian objectives of the framework program and the traditional framing of civil and defense-related issues (Åhlén, interview, 1996, and interview, officials from DG XII 1997; see also Mörth 1998, 2000a). Consequently, the status quo prevailed. However, it was still possible to finance dual-use technology indirectly in the program. Defense–related organizations can already participate in research, provided, of course, that they comply with the civilian objectives and rules of the programs. In addition, the key actions supported in the proposal were closely linked to technology with extensive military applications, especially aeronautics. Aeronautics is in the thematic program, "Competitive and sustainable growth." This program receives 2705 Million ECU (1998-2002), out of which 700 are allocated for a key action, "New perspectives for aeronautics." This is quite a large sum; "Global change" in the fifth framework program, for instance, only gets 301 Million ECU (Huusela, interview 1997). So, in a way it was business as usual. However, the issue of the framework program and dual-use technology was only part of an emerging policy from the Commission that aimed at breaking down the differences between RTD and industrial activities within the civilian-and defense-related spheres. In the era of information technology and technological globalization, this difference was perceived as obsolete.

The communication from January 1996 showed the Commission's ambition to pursue a more comprehensive industrial policy—an action plan—that included not only Europe's civilian industry, but also its defense–related industry. Although the future of the defense industry had been discussed earlier, the communication was the first comprehensive document from the Commission on the problems of the European defense industry. This was a rather bold initiative due to the fact that this sector has been regarded as an area of exclusive national prerogative. Article 296 (formerly Article 223) in the Treaty on European Union allows govern–ments to exempt defense firms from EU rules on mergers, monopolies, and procurement. The article states ,

1. The provisions of this Treaty shall not preclude the application of the following rules: no Member State shall be obliged to supply information the disclosure of which it considers contrary to the essential interests of its security; any Member State may take such measures as it considers necessary for the protection of its security which are connected with the production of trade in arms, munitions and war materiel; such measures shall not adversely affect the conditions of competition in the common market regarding products which are not identified for specifically military purposes.
2. The Council may, acting unanimously on a proposal from the Commission, make changes to the list, which it drew up on 15 April 1958, of the products to which the provisions of paragraph 1 (b) apply.[6]

The opinion of the Commission was that "hitherto Article 223 of the EC Treaty has placed limits on the Community framework by allowing exemptions from the provisions of the Treaty for 'production of or trade in arms, munitions and war materiel'" (Com 96 (10), 14). This exemption applies only under particular circumstances and conditions since the same article adds that the national measures on the subject must be "necessary for the protection of the essential interests of the security" of the member states and must "not adversely affect the conditions of competition in the common market regarding products which are not intended for specifically military purposes" (COM, (96), 10, 14). According to the Commission, some member states have interpreted this article broadly with the result that the EU industry has lost ground to the U.S. industry (COM, (96), 10, 14 see also Carvalho 2000). "On an overall industry level the aforementioned trade figures give a strong indication that the European defense-related industry has experienced a worsening of its competitive position vis-à-vis the U.S. industry since the 1980s" (COM (96) 10, 7). A stricter interpretation of Article 296, and other actions to facilitate integration of defense-related industrial activities, "will have to take account not only of the specific nature of the armaments sector but also of its essential and ever closer links with the civil sector (dual-use technologies, components, products and production installations) in order to encourage the development of technological and industrial synergies between those two sectors at the European level" (COM (96) 10, 7). Already in 1990 the European Commission discussed the future of Article 296 and claimed that "it is also in the interest of the Community to bring armaments protection and trade fully under the discipline of the common market, which would involve inter alia the removal of 223" (COM (90), 600 final, 5).

The question of the abolition of Article 296, or a stricter interpretation of the article, is, of course, politically sensitive since this would further bring armaments under the regulations of the internal market. At present, Article 296 may be invoked by member states in their defense deals, but all deals would normally be considered within the market rules by the Commission for possible overlaps that might raise antitrust concerns. However, so far the Commission has never opposed a defense industrial merger nor had the Court try to determine whether a country has broken the rules in not notifying a merger within the defense industry (Interview, senior official DG IV, September 1999). This could change, of course, if the defense industry should start to be treated as any other industrial sector. In a scenario in which mergers, joint ventures, etc. among the European industries will increase, it has also been argued that it will become more difficult in practice to judge whether Article 296 is applicable (Interview, senior official DG IV, September 1999).

Although the Commission presented its communication on the restructuring of the European defense industry as a unified proposal, the Directorate General on Industrial Affairs heavily influenced the COM document[7] (Interviews, senior officials in DG I, February 1997, IA, February 1997, September 1999, III, February 1997, June 1997, May 1998, September 1999, and IV, September 1999). It is clear that DG III pursued a policy of introducing more industrial policy into the defense industry, meaning that the rules of the internal market could, after necessary adaptations, be used for this industry as well. One important industrial sector in the ongoing discussion of the future of the defense industry is that of aerospace. DG III had close relations with the industry, both bilaterally and with EDIG and AECMA.[8] Various efforts have been made to create one unified voice within the European aerospace industry, but the obstacles to European aerospace cooperation were for a long time multiple and complex. To create a European defense industrial strategy there have been several meetings between DG III and EDIG on the issue of armaments (Interviews, officials at DG III and the General Secretary of EDIG, 1999). EDIG has prepared a number of position papers on intra-Community transfer, taxation, European Company Statutes, and other defense industrial issues. These position papers have been discussed with the "representatives of a number of Directorates" (EDIG 1997, 2). The Commission and EDIG agree on the need to create a collective effort "to tackle the various sensitive issues leading to the establishment of a European Armaments Market where Defence Industry will survive to remain competitive and capable of catering for the European Armaments needs" (EDIG 1995, 2).

According to DG III, the incorporation of the defense industry into the first pillar must be implemented systematically. Although DG III recognizes that the defense industry is a very special market, with its close relationship with the state, and that it therefore differs from other sectors of the economy, it is also obvious that it can be approached from a perspective of cost effectiveness. This is also its task within the Commission. In fall 1997, DG III presented a "Draft Action Plan for the Defence-Related Industry" outlining measures for the short term as well as the long term.[9] A first step is to begin a process of standardization of European armaments, intended to rationalize the different sets of standards currently being used by the defense ministries of the member states. This process of standardization also entails common rules of public procurement. In a longer perspective, this standardization process must also extend to differing national export policies with regard to conventional arms (see below). The next step would be to incorporate the defense industry sector into the EU's competition policy and state-aid regulations. During this stage there would also be a need for a European Armaments Agency in charge of conducting programs for cooperation and R&D related to armaments.

Clearly, the profile of DG III and its then-commissioner, Martin Bangemann, was very high concerning defense industry issues. They organized seminars, informal meetings with representatives of the industries, national administrators, and other European bodies (Mörth 1998; 2000a). At a conference in June 1996—"The Future of Europe's Defence Industry"— Bangemann urged the national administrators to take part in the "decision-finding process" that the Commission had initiated, adding that they should tell "us what are the main problems and priorities" (Bangemann, 1996) The commissioner emphasized the problems that the European defense industry was facing with respect to the United States. "First, if European defence industry will not overcome its structural problems we may lose the capacity either to compete efficiently or to co-operate on equal terms with the USA. Secondly, we will be challenged with technology gaps and a disappearance of technological skills. As nowadays, dual-use technologies are widespread, this may also have impacts in the commercial sphere. Given the long period of time for developing new technologies and new systems, these gaps could not be filled in a few years. This would have substantial economic and political consequences" (Bangemann 1996).

We can thus conclude that the activity of the Commission, especially DG III and EDIG, was intense during the early to mid-1990s on the future of the European defense industry. In November 1997 the Commission presented another communication on the defense industry, and in that communication the market frame is less salient than in the 1996 communication. In fact, in the communication the Commission argues that the issue of armaments must be conceptualized from both a market and a defense frame. This communication is analyzed in chapter 5 since it is an important component in the emerging common field on the issue of armaments.

An interesting question is how the activity from the Commission was politically received.[10] The political reaction toward this effort to strengthen defense industrial competitiveness did not come until 1997 (see below), and it was not an initiative within the EU. The communication was, however, discussed in the European Parliament, and in May the so-called Titely report was accepted in plenary session.[11] The Parliament declared that a common policy on armaments "will ultimately require Article 223 to be revised" (A4-0076/97: 8). The resolution supports the Commission's diagnosis of the situation for the defense industry and its observation that it is difficult to distinguish between defense-related RTD and civilian research projects.[12] However, in an early response from the Committee on Research, Technological Development, and Energy, it questioned the idea of financing the defense sector by civilian RTD money since this could compromise the civilian objectives (European Parliament, The Committee on Research, Technological Development and Energy, 1996). The civilian orientation of the framework program for RTD is emphasized in the resolution, but it is also stated that there should be better coordination between defense–oriented and civilian RTD programs "so as to avoid duplication of effort" (European Parliament, The Committee on Research, Technological Development and Energy, 1996, paragraph 27).

The European Parliament has been rather active on the issue of armaments and the European defense industry. The Commission's and the Parliament's involvement in the defense industry have been through the EU's industrial and regional policies, especially the regional program KONVER (helps regions heavily dependent on defense sector cope with consequences of cuts in defense budgets), which was completed in 1999. Another specific program to support regions suffering from loss of employment in the defense industry sector was initiated in 1991 by the European Parliament (Regions Périphériques et Activités Fragiles, Öberg 1992). The legal base for the industrial activities was given in the Single European Act. Article 130f was intended to strengthen the scientific and technological basis for European industry. The European Parliament has also initiated the European network ADRIANE (the Aerospace and Defense Regional Initiative and Network in Europe). The network embraced all sectors involved in aerospace and defense work with a special focus on small and medium sized enterprises and labor market issues related to the defense industry. The emphasis on employment and the defense industry was quite new. The Commission's reports rather seldom discussed the labor market consequences of national restructuring of its defense industry. However, in September 1999 the ADRIANE initiative seemed to have reached a deadlock (Gunilla Carlsson, interview, 1999). One explanation for this deadlock is that the network stopped being active (at least officially) when the Cologne process on the EU's foreign, security, and defense policy was activated (see chapter 3).

To sum up, the Commission has been very active in framing the issue of armaments as belonging to the market field. The political reactions have been rather limited. Chapter 3 showed that the issue has been the object of political attention, but not inside the EU, at least not until very recently. We now turn our attention to the defense industry, especially to the aerospace industry in Europe, and its activities during the 1990s and in the early part of this century.

The Creation of a European Aerospace and Defence Company (EADC)

There have been major changes in the European industrial landscape during the 1990s. National and state–owned companies—national champions—have been transformed into private European and transatlantic companies.[13] The transnational linkages between the companies are multiple and complex (for an overview see Schmitt 2000b; Axelson and James 2000). Major industrial changes took place in 1999, which, of course, could be explained by the general development in the European defense policy process (chapter 3; see also Heisbourg 2000). It is difficult to imagine that complex cross–mergers between defense industries would have taken place if there had been no credible political development of a European defense policy. This being said, it is also quite clear that the industry itself created pressure for political initiatives and also that the process toward a strong European defense industry is driven by market factors and not only by the logic within the defense field. The aim in this part of the chapter is to discuss the industrial and the political idea of creating a European aerospace company—the European Aerospace and Defence Company (EADC). The general rationale behind the importance of creating a strong European defense industry is partly outlined in the previous section, namely, that the borders between technologies are considered to be blurred according to the Commission, and that the national protection of the defense industry is an obstacle to a common European effort to strengthen the region's industrial and RTD base.

Many observers within industry and the European Commission claim that the European aerospace industry suffers from the increasingly acute effects of the continued partitioning of its industrial structures. "Despite enormous efforts, at both national and European levels, the European aircraft industry is still suffering from the effects of excessively long partitioning of its industrial structures, particularly in the equipment sector. To adapt the industrial structures to the internal market and to the increasing globalization of the economy, the priority questions are: do current competition conditions allow the European aeronautical industry to be effective? What suitable measures can be taken to improve European competitiveness?" (COM (92) 164: 1c) This diagnosis of the aerospace industry is presented in several communications from the Commission during the 1990s. "National markets can no longer provide a sufficiently strong base to support a full-range independent aerospace activity" (COM (96), 617 Final, 6). In contrast to the situation in the United States, European industry is fragmented because of national boundaries and separate research and defense policies (COM (96), 617 Final). The United States is the market leader in both civilian and military aerospace, and the recent history of U.S. industry has been one of consolidation (James 1999).

In 1993, the U.S. defense secretary at the time, Les Aspin, invited the defense industry for a "last supper" at which he made it clear that the industrial defense companies had to restructure (James 1999). This was more than a rhetoric change by the U.S. government; it was followed by a merger wave in the American defense industry—from twenty companies to three prime suppliers: Boeing–McDonnellDouglas, Lockheed Martin, and Raytheon (James 1999). The European defense industry, as the European Commission, closely monitored the American defense industrial consolidation and its "new R&D-centered drive towards a 'post Cold War military–technological paradigm'" (Lovering 2000, 12). One major turning point from the European defense and civil aerospace industry perspective was when Boeing and McDonnellDouglas merged in 1997. Until then, the civil European aircraft project, Airbus, had challenged Boeing, but after the American merger the power balance between the American and European companies profoundly changed (Gissler, 2000). Thus, the emergence of American defense industry giants "triggered a scramble to consolidate in Europe in an effort to create firms of a size thought necessary to compete in the new industry environment and to allow Europe to enter into a relationship of equals with the U.S. giants" (Axelson and James 2000, 35).

The comparison with the situation in the United States is striking in the reports from the Commission. "In comparison with the US, the pace of consolidation in Europe has been dramatically slow . . . in recent years the European aerospace industry has lost ground to that of the restructured and revitalised US industry" (COM (97) 466, 5-6). Indeed, Europe's consolidation of its defense industry must include measures to build in mechanisms that will prevent American companies from buying European companies (Interview, senior official, DG III, September 1999). In a report from a working party from CEPS, the Brussels-based think tank, consisting of representatives from the industry and from the EU, the WEU, and NATO—it is stated that the "European defence industry suffers from a widely fragmented home market base, overcapacity and duplication of official procedures and processes. Its ability to compete in the global armaments market is threatened by these impediments" (CEPS 1999, 22). What is needed is thus "rationalization" and "consolidation" of the sector and the creation of European companies (CEPS 1999, 22).

The problem with a fragmented structure of the European aerospace industry is also put forward by the industry, especially by its European branch organization, AECMA, and the general European branch organization of the defense industry, EDIG.[14] "Only if the current fragmented structure of the industry is superseded by a few large transnational organisations which are unimpeded by individual policies based on national borders can the industry remain both competitive and profitable" (AECMA 1996, 55). Although the market for aerospace products is global, "Europe requires an aerospace industry to support its role in global air transport, to allow an autonomous foreign policy and defence position, to safeguard its access to space and to contribute to technologically driven growth" (AECMA 1996, 23). This means that a favored way for the restructuring of the European aerospace industry "is to replace the current loose cooperative arrangements with transnational company structures which have truly European dimensions" (AECMA 1996, 23). The industrial branch organizations thus present rather clear strategies and recommendations to be addressed by the decision makers. "The major prerequisite to establish–ing a European Domestic Market will be that the member Governments harmonize their operational requirements to enable common procurements. Other sensitive and complicated problems will have to resolve: harmonization of acquisition procedures, laws and regulations, standards, security of supply, reciprocity of market access within Europe, work share and industrial return and export controls" (EDIG 1995, 3, see also 14). Europe should not "become dependent on third countries in the area of armaments" (EDIG 1995, 2). Individual industrialists seldom make any statements on European strategies for defense industries. They also tend to downplay their own role in forming a European regulatory framework for armaments. One illustration of this is the statement by the senior executive vice president of Saab, Lars Josefsson: "It is important to remember that the playing field and the rules for the defence industry is purely political as the rules are set by the national Governments" (Josefsson 2000).

The technology and capability gap between the Unitd States and Europe is also presented by several journals specializing in defense industry matters, for instance *Aviation Week & Space Technology, Defense News,* and *Jane's Defence Weekly* (*JDW*). Comparisons between the United States and Europe, for instance, concerning defence budgets or RTD budgets, always show that the United States spends more. Implicitly, and sometimes even explicitly, these journals present the model for Europe—that of the United States. The journals also underline the image of a war between the United States and Europe on armaments matters (see for instance the case of Meteor, Morocco, 2000a).

It should be noted that there are also, although they are in the minority, other assessments of the diagnosis of the European defense industry in relation to that of the United States. The researcher Michael Brzoska (at the Bonn International Center for Conversion), for instance, argues that Europe "is disadvantaged in fields where economies of scale matter, such as fighter aircraft production. However it is better off in fields where they are of lesser importance, such as shipbuilding" (Brzoska 2000, 8). This nuanced image of the lagging-behind theme is also presented by Hagelin, a researcher at the Stockholm International Peace Research Institute (Hagelin 1999).

So, the general image of European defense and aerospace is one of fragmentation, and one in which the industry needs to be consolidated in order to compete with American companies. In December 1997 the French, German, and U.K. governments issued a statement in which they stressed the vital political and economic interest of restructuring the European defense industry. The political initiative has been interpreted as a reaction toward the industry and its demand for political activities to enhance the creation of transnational defense companies (Interviews, officials, DG IA, May 1998, September 1999, and DG III May 1998, September 1999, representatives from SAAB February 1998, Törnblom, interview, 1998, Woodcock, interview, 1999). Until then, the European governments had been rather silent on the question of the future of the European defense industry, and the most active actor at the European level was the Commission. In summer 1998 the three governments plus the governments in Sweden, Spain, and Italy presented the Letter of Intent initiative (LOI), which aimed to enhance the creation of transnational defense companies (chapter 5). The statement in November 1997 was thus the preliminary phase of this political initiative and agreement. In that statement the three governments made it clear that they wanted to launch various measures to enhance transnational industrial collaboration, and in the declaration the governments urged the national champions to present a plan and timetable for industrial restructuring and integration (Schmitt 2000c). Paradoxically,

the political vision at the time was focused on the creation of *one* company, a European Aerospace and Defence Company, although this could be
questioned from a market and competitiveness perspective. The company
would include the Airbus partners (France—Aérospatiale, Germany—
DASA, United Kingdom—British Aerospace, Spain—Casa) and Italy's
Finmeccania–Alenia and Sweden's Saab.[15] There were several rounds of
discussion regarding the makeup of this company, in the industry as well
as in political circles. The hurdles were many. One stumbling block was
the general lack of a European company statute. The Commission stated
in a communication in 1992 that with the European company statute, "the
Community will provide the European aircraft industry with a legal framework for adapting its industrial and legal structures to the conditions of
global competition" (COM, (92), 164 Final, 24).[16]

The question of European companies, or *societas europea*, goes back
to the late 1950s, and it was not until the Intergovernmental Conference
in Nice in December 2000 that the heads of state could agree on the statute.[17] Thus, the question of a European company statute affects not only
the defense industry, but also other industrial sectors. National differences
concerning social and legal conditions and tax differences have, however, prevented the adoption of a supranational legal form.[18] The creation
of a European company statute would require deregulation of national
company rules. It would also give rise to pressure for tax harmonization.
Today, companies can take advantage of different tax rates in the European
countries. A European company statute would entail a decrease in national
disputes between partners in joint ventures since the company would be
supranational. The creation of a statute was supported by the defense industry. "The legal and regulatory framework for the creation of aerospace
industrial structure in European dimensions has to improve. In addition to
regulatory acceptance of mergers, a European company statute is required" (AECMA 1996, 54). "It is recommended that the Commission pursues the establishment of the necessary Directives and regulations that will
enable the establishment of SEs as soon as is practicable" (EDIG 1996).[19]
Indeed, a common legal framework for companies at the European level is
regarded as important (Ivarsson, interview, 1999). One advantage of such
a reform is that multinational firms would cut their administrative costs by
setting up as EU-wide companies instead of fifteen national firms (Coss
2000) . Today the companies create holding companies in the country with
the lowest tax rates (Ivarsson, interview 1999). In the report from CEPS it
is stated that there is a need for a European company statute to create an

EADC (CEPS 1999). The industry also argued that tax and labor law "as well as social regulations in Europe should be brought to a higher degree of convergence. Adoption of a single European currency would also help" (AECMA 1996, 54).

Another obstacle to the creation of the EADC, and a more important one, was the fact that the defense industry in Europe was heterogeneous and that the relationships between the state and industry in various national settings varied. However, in March 1998 the would-be partners in the EADC presented a report on a possible map for the implementation of the European company. The report was classified, but it was quite obvious that it was a difficult task to present a blueprint for a European company (CEPS 1999; Schmitt 2000c). "It therefore set no timetable for an eventual merger and studies continued" (CEPS 1999, 49). A second report from the industry was presented in mid-November 1998 (Schmitt 2000c). However, the discussions between the governments and the industry "never reached the stage of real negotiations. They were essentially an exchange of ideas and a general discussion of possible avenues to explore" (Schmitt 2000c, 30). One key barrier was the differing shareholding structures among Europe's defense and aerospace players. This was especially the case with France, in which defense industrial companies had a mix of state and private ownership. Thus, one important obstacle to the EADC's creation was the reluctance of the French government to part completely with its stake in Aérospatiale, the likely French partner in the combined group. BAE[20] and DASA[21] have been against any state ownership. However, in February 1999 the French government announced the terms of the privatization of Aérospatiale and its merger with the Matra offshoot of Lagardère. The new company, Aérospatiale Matra, is France's biggest aerospace producer (Britz 2000). The French government's reluctance to privatize its defense industry seems to have functioned as a convenient excuse for not moving fast enough toward creating a European company.

Rather soon afterward, the industrial development took a step back from the creation of one company. In January 1999, British Aerospace confirmed that it would buy General Electric Company's Marconi Electronics, creating the third-largest defense and aerospace group in the world. Germany's DASA warned that the creation of a pan-European aerospace and defense company might be postponed indefinitely (*Financial Times* 25 January 1999). Prior to the takeover it was often mentioned in the specialized press that the best way to push forward plans for a consolidated European Aerospace and Defence Company was through a link-up between

BAE and DASA. One consequence of the deal between BAE and Marconi was that the balance between BAE and DASA changed, which meant that the willingness of DASA to form a joint company with BAE decreased. It was therefore argued that the vision of a large European defense company was unrealistic (*FT* 25 January 1999). In October 1999, a merger between Aérospatiale Matra and DASA was announced. The new company, which would form a military aviation joint venture, was named EADS (European Aeronautic Defence and Space Company), not to be confused with EADC. The new holding company was officially founded in July 2000. It will be registered in the Netherlands and have dual headquarters in Munich and Paris (*FT* 15 October 1999). The company will also include CASA, which has agreed to merge with DASA (*FT* October 1999). Furthermore, in December 1999 it was announced that the Italian company Finmeccanica had received a formal offer from EADS. "The move comes weeks after BAE showed similar interest in forging an alliance with Finmeccanica's Alenia Aerospazio subsidiary, especially in the area of combat aircraft" (*FT* 30 December 1999). Indeed, both BAE Systems (formerly BAE) and EADS are competing to strike an alliance with the Italian state-owned Finmeccanica in order to dominate the European military aircraft industry (Blitz and Nicoll 2000). In April 2000 the Italian state-controlled defense and industrial group, Finmeccanica, announced that it had selected EADS as the alliance partner for its military and civil aircraft unit (*FT* 14 April 2000). "With the addition of CASA, EADS has over 95,000 employees and a turnover of GBP 21 billion" (Schmitt 2000c, 39).

What is interesting with the new company (EADS) is the fact that "for the first time, national champions are merging all their assets (with the exception of DASA's aero-engine subsidiary MTU, which will be retained in the DaimlerChrysler group). With the recent inclusion of CASA, what began last October as a Franco-German rapprochement is turning into a truly European grouping" (Schmitt 2000a, 4). The new European grouping will enjoy a dominant position in the European aerospace industry. The group will, for instance, hold 80 percent of Airbus[22] (which leaves 20 percent for BAE Systems), 45.76 percent of Dassault Aviation and 43 percent of Eurofighter (BAE Systems holds 37.5 percent) (Schmitt 2000a, Schmitt 2000c). Finmeccanica's decision to join forces with EADS means that the combined grouping has a 62.5 percent equity stake in Eurofighter[23] (the U.K. group controls close to 45 percent of the value of the project, *FT* 14 April 2000). The agreement between EADS and Finmeccanica also paves the way for the Italian company to join the Airbus civil aircraft consortium for the first time (*FT* 14 April 2000). In May 2000 the European Commission cleared the creation of the new company. Even though the companies involved have overlapping interests, the political pressure on the Commission from the governments concerned is very high (Cordes 2000).

Thus, for now, Europe's aerospace and defense industry is divided into two large industrial groups in the aerospace sector: BAE Systems (formerly BAE) and EADS Aerospatiale/DASA/CASA and Finmeccanica Alenia.[24] There are links between the two large European groupings through participation in the main European programs, such as Airbus and Eurofighter (Schmitt 2000c). They are partners and competitors at the same time. Furthermore, the organizations of the companies differ. One one hand, BAE Systems is "vertically integrated and highly specialised in defence; on the other EADS, horizontally integrated and strong in the civil sector" (Schmitt 2000a, 4). In contrast to BAE Systems, EADS consists of a rather complex industry group (a private group, a temporarily state-owned company, and a partially privatized group). The formation of EADS and its recent agreement with Finmeccanica has, however, been interpreted as a strategy toward free enterprise. Politically driven companies are no longer tolerated in a competition-driven global market, according to Henri Martre, former head of DGA French Armaments Agency (Sparaco 2000a).

The notion of one big European company, EADC, thus seems dead, at least for the present. The idea has been questioned from an industrial competitiveness perspective and as a symbol for fortress Europe. "We have the capability but in doing that it is essential that we do not establish fortress Europe by creating a single acquisition agency and only one international industry in each industrial area. That would merely lead to planned economy" (Josefsson 2000). One of my interviewees in the Commission said in September 1999 that very few believed in the creation of *one* company. The notion of EADC was instead an effective way to provoke and thus generate action to get rid of various barriers. Even though the establishment of one company had been politically and industrially desirable, it would have been against the EU's competition law and policy. In August 1999, the chief executive of BAE, John Weston, said in a newspaper article after the BAE purchase of Marconi that "if you look at it, the same sort if idea we had before in EADC cannot be realized anymore, unless you split BAE. So we probably have to develop new (consolidation) ideas, but that doesn't mean the end of restructuring in Europe. Not at all, but it probably will move in a different direction" (Muradian 1999). In March 2000

Manfred Bischop, chief executive of DASA (who became cochairman of EADS), said "the EADS partners were confident that agreement could also be reached with BAE Systems of the UK" (*FT* 9 March 2000).

From a European political perspective, the question of what kind of linkages the European companies will have with American companies is, of course, crucial. In August 1999 John Weston stated, "We never conceived EADC as something that represented Fortress Europe, with everything in it. We actually saw it as a mechanism for how to get some of these deals together across the national boundaries of Europe" (Muradian 1999). In fact, the companies are mixed up with each other already (Woodcock, interview, 1999). Interestingly, in February 2000 the French coprésident exécutif of EADS, Philippe Camus, stated that EADS would not constitute a fortress Europe. "Au contraire, il constitue une formidable occasion pour engager des discussions d'egal a egal avec nous conurrents americains. Je suis très en faveur d'une cooperation transatlantique" (Jakubyszyn and Rocco, 2000). The transatlantic linkages already exist. EADS and Northrop Grumman "are evaluating potential business alliances and have already agreed to offer a 'European version' of a Northrop Grumman developed weather and navigation radar for the Airbus A400M military transport aircraft" (Axelson and James 2000, 45). Furthermore, in October 1999 it was announced that Marconi Electronic Systems (owned by BAE), Aérospatiale Matra and DaimlerChrysler Aerospace agreed to merge their space activities into a new company, Astrium (*Aviation Week & Space Technology*, 25 October 1999; 5 June 2000). The same month it was also announced that Aérospatiale Matra, British Aerospace and Italy's Finmeccanica were about to merge their missile businesses (*Aviation Week & Space Technology*, 25 October 1999). If Finmeccanica enters into Airbus, a transatlantic problem will be created. This is because Finmeccanica already has a close relationship with Boeing—the arch-rival of Airbus (*FT* 14 April 2000).

In December 1999 it was stated that BAE Systems of the United Kingdom, and Boeing, the U.S. aircraft and defense manufacturer, were engaged in exploratory talks that could result in a merger of their defense businesses (Nicoll 1999). BAE and Boeing are already involved in a number of collaborative defense-related projects, and the two governments have stressed the potential benefits of defense cooperation between the United Kingdom and the United States (Nicoll 1999). Indeed, one scenario suggests that in about five years BAE's linkages with Boeing will intensify and that the American company Lockheed Martin "could become the

American friend of EADS" (*The Economist*, 23 December 2000, 112). BAE also has a long-standing partnership with Lockheed Martin. In November 2000 BAE bought parts of Lockheed Martin, making it one of the world's largest defense groups (*The Economist*, 23 December 2000).

The merger between British Aerospace and General Electric Company's defense division illustrated the search for global access, particularly to the large U.S. defense market. The combined new company will do 25 percent of its business in the United States. The logic that makes the Atlantic option attractive is, of course, compelling. The military hardware budgets have been decreasing on both sides of the Atlantic. The production of high technology, which is so important for the military, is dependent on global and commercial business. Hence, a market-driven consolidation tends to lead toward closer transatlantic relations. Politically, however, governments seem to be rather reluctant to move toward a more market-driven and transatlantic consolidation (chapter 5). One political problem is the uncertainty regarding the building of ESDI, which is what the eventual role of the United States and NATO may be (chapter 4). From an industrial and business community perspective it could be argued that the political polarization between the United States and Europe will be (or perhaps is already) less relevant among the defense companies. It is therefore likely that in the long run there will be U.S. firms "that become European and European firms that become American, within and across product lines, enhancing companies and efficiency in a broader transatlantic market" (Adams 1999). From the European Commission's perspective, however, it is very important to create a strong Europe in relation to the United States. Indeed, the European consolidation of the European defense industry must include measures to build in mechanisms that will prevent American companies from buying European companies (Interview, senior official, DG III, September 1999).

An illustration of the competition between Europe and the United States is the arming of the Eurofighter, which is a multinational fighter project. Whether the European countries choose the missile manufactured by Raytheon or one produced by the European Meteor consortium has been seen as an indication of whether U.S. fears of a "fortress Europe" may be realized. Indeed, the decision by the governments in the United Kingdom, Germany, Spain. and Italy to select Meteor is seen as crucial in the ongoing consolidation of the aerospace industry (*FT* 18 February 2000). From a European industrial as well as political perspective, Meteor is primarily about breaking a U.S. monopoly in the missile sector. The

European consortium consists of a joint venture by Aérospatiale Matra and BAE Systems and the American company Boeing. In February of 2000 the deputy chief director of Matra BAE, Alan Garwood, declared that this cooperation provided "a very clear signal to the U.S. government that strong relationships with American industry are very important to us" (*FT* 18 February 2000). In May 2000 the British government decided it would buy Meteor air-to-air missiles and not the American Amraam missile (*FT* 17 May 2000). "For the first time Europe will equip its fighter aircraft with a European air-to-air missile, creating interoperability and independence to export" (Moroccoo 2000a).

To sum up, the European defense industry is traversing a period of dramatic change. The earlier vision of one European aerospace company is dead. Instead, Europe is doing what the United States did earlier, namely, creating a small number of large companies able to compete with the U.S. giants.

Concluding Analysis

When the issue of armaments is framed in the market field, a different set of questions and actors is activated than is the case with the defense field. The defense aspects of the issue are less salient. Instead, the overall question concerns how to create an internal market for armaments and thus how to reduce national protection of the defense industry. The Commission's external environment is interpreted from an industrial/technological and market perspective and the threats to Europe are of a civilian nature. The market field means determining how an issue is diagnosed (that is, problems of technological and economic competitiveness toward the United States) and what has to be done (that is, act at the European level in the community pillar). Indeed, the market field is based on the notion of Europe's technological and economic decline in relation to the United States.

Parts of the European Commission and the industry have been working for several years toward the same goal—the creation of a closer European cooperation on armaments. Like a Trojan horse, the Commission has pursued a market frame to gain influence over an issue that has traditionally been framed as a defense and security issue. Article 296 of the Amsterdam Treaty excludes armaments from the internal market due to national security interests. However, the Commission, together with the industry, has special journals, and policy centers such as CEPS have created a political crisis awareness pointing to various economic, industrial, and technological threats from the United States. An image of Europe in disarray is presented that legitimizes a stricter interpretation of Article 296 and the protection of national security interests. Important values are at stake, which makes it necessary to think in European terms.

The Commission's initiatives on the future of the European defense industry, especially the communication from January 1996, did not arrive out of the blue. The communications reactivated the classic lagging-behind theme between Europe and the United States and followed a well-trodden path in the European integration process. Through the market framing activity—contrasting Europe with the United States—a sense of European unity and identity is created. Thus, framing is not only about strategic interests, it also constitutes the identity of actors. A sense of European actorness is created. The external activity becomes an important part of the internal organizing of the European Union and will influence the appropriate policy response. The European cohesion process on industrial and technological issues became stronger in the 1980s, when several European RTD programs were launched that created a European technological community. With the Single European Act and the Maastricht Treaty, the legal base for a more coordinated and European industrial policy was strengthened. Thus, the ongoing liberalization of the national defense industries is part of this liberalization trend in the European Union, which started with the creation of the internal market and the national deregulation of the civilian high-technology industries (Sandholtz 1992). The empirical analysis in this chapter shows that the European Commission has now transferred this reasoning on the importance of knowledge-based power and the view of the adversary to the European defense industrial context. This time, the market integration entails the gradual harmonization of standards, or at least the creation of a framework within which a standardization process could take place. The aim is to create a body of common rules, which will lead to the emergence of common standards within the defense industrial sector.

It is quite clear that the European Commission has created room for maneuver within this issue and that it has tried to organize a European experience. The early initiatives by the Commission were a convenient way for the EU member governments to discuss the question without needing to enter politically dangerous territory (that is, defense policy). The issue of armaments was thus put on the political agenda by the Commission, not by the governments, and this was done in a way that would not jeopardize the defense process. Furthermore, activity within the defense industries in Europe suggests that the increased industrial collaboration into transnational defense companies has put pressure on the governments to "act" European instead of national. The only item on the political agenda seemed to be to help industry cope in a more internationalized and competitive environment. The governments were, however, against a communitarization (supranational decision making) of the issue of armaments as suggested by the Commission.

In comparison to how the issue of armaments has been part of the two fields—defense and market—the empirical evidence shows that the political breakthrough in the EU for the issue occurred when the EU became involved with defense issues. Until then, the issue of armaments had been part of NATO and the WEU but was not within the second pillar of EU. This chapter shows that the issue of armaments was part of the EU for several years but that it was part of a market field and the civilian path of the European integration process. Thus, when the governments activated the issue of armaments in the defense field, they codified a political and industrial process that was already under way in the market field.

Table 1 condenses the empirical analysis in chapters 3 and 4. The table shows how the two fields belong to different political and ideational contexts in European politics. The fields activate different questions and actors; they do so because they are based on different regulative and constitutive rules. The regulative rules are various articles within the first pillar (the market field) and the second pillar (the defense field) that concern market and CFSP questions. The Commission is a strong actor in the first pillar, whereas the European Council and the Council are the crucial actors in the second pillar. According to the European Council and the Council, a closer cooperation on armaments should be organized as an intergovernmental cooperation with a weak role for the Commission. The proponents of the market field argue, on the other hand, that the supranational regulatory framework in the community pillar is relevant.

The constitutive rule in the defense field is the general notion that defense issues should be dealt with in an intergovernmental way since these issues concern national sovereignty. Traditional security, that is security expressed in terms of military threats and power, is based on the logic of anarchy, which necessitates that the states control defense issues. The political ambition to create a European capacity to handle military crises—and to create a European actorness on defense—is combined with an intergovernmental decision-making process. The driving force behind the need to form such an actor capacity is the changing security situation after the end of the Cold War.

Table 4.1. Two Fields and the Issue of Armaments

Defense field—Anarchy and intergovernmental cooperation	**Market field**—Interdependence and supranational cooperation
Dynamics: end of the Cold War, military interoperability	Dynamics: internationalization of high-tech industry, internal market
Prime issues: Petersberg tasks, CJTF, PFP, European Security and Defense Identity (ESDI),[25] European Security and Defense Policy (ESDP)	Prime issue: A European armaments market, European companies
Prime actors: NATO, WEU, EU (Second Pillar)	Prime actors: EU (First Pillar), Industry

The constitutive rule in the market field is that issues within the first pillar should be handled according to a supranational decision-making process since these issues concern Europe's economic and technological competitiveness vis-à-vis the United States. The underlying logic behind this rule is the increasing economic and technological interdependence between states that necessitates thinking in European terms. The dynamics behind the need to form strong European defense companies, and to create a European armaments market, is thus to be found in the ongoing technological and industrial internationalization of technology. The prime issue within the market field is to strengthen Europe's technological and economic capacity.

To sum up, organizations identify themselves with different projects related to European integration—the political economy project or the defense and security project. To put it simply, where you stand depends on where you sit. Traditionally, these two organizational fields have been living separate lives with very little communication and interaction between them. This is so because the EU on the one hand, and the WEU and NATO, on the other, are built on different logics in the international system. The European Union is built on the logic of economic interdependence and on the notion of a strong European market. The WEU and NATO are built on the logic of anarchy and on the notion of enemies. In markets, however, there are no enemies, only competitors. Indeed, "in a world of economic globalism in which production becomes decoupled from individual states, it becomes more and more difficult to constitute an Other that might be transformed into a threatening enemy" (Lipschutz, 1995, 220).

This does not mean that an organization from the defense field cannot or will not discuss the issue of armaments from a market perspective. The major concern on the issue of armaments will, however, be dominated by defense or market concerns depending on the field to which the organization belongs. This also means that an organization within the defense field is most unlikely to favor a supranational decision–making structure even though it has considered market aspects of the issue of armaments. Hence, a market organization can take into account the fact that defense considerations are important but that the political and legal measures are market oriented.

There are similarities between the market and defense fields in the sense that they both concern the notion of Europe's need to enhance its capacity in relation to that of the United States. What is presented in these two fields is the need to create a European actor and to pursue European economic and military strategies. The Other in this identity-seeking and capacity-building process is clearly the United States. However, the notion of the United States as the enemy in the global race for competitiveness is not found in the defense field. Paradoxically, it is the defense field that presents a diffuse image of enemies and threats whereas the market field presents a rather clear diagnosis and solutions to the problems that are facing Europe. The conventional wisdom of the two logics in the international system, anarchy and interdependence, would presume that the anarchy rhetoric would dominate the defense field and that the rhetoric on interdependence would be predominant in the market field (see also chapter 3). The lack of a clearly defined enemy in the defense field can, of course, be explained by the end of the Cold War and the fundamental changes in the European security architecture. What is interesting, however, is that the traditional security rhetoric of power balance and anarchy is being diffused to the European discourse on market-making activities.

The empirical analysis in chapters 3 and 4 has shown that to frame an issue is to assign meaning to it and that frames make a difference in how an issue is politically handled. By giving the issue a special meaning—as a technological/industrial deficit vis-à-vis the United States or as a question of a European defense capacity—history is written; causal linkages are established and guidance for future action is presented. We now turn to an analysis of how these two frames and fields have moved toward each other. The next chapter analyzes this transformation of the two rather solid and independent historically developed fields.

Notes

1. The expression "Three waves of Technology-Gap fever" is from Sandholtz, 1992.
2. The new number is 308 according to the consolidated treaty after Amsterdam. The article states, "If action by the Community should prove necessary to attain, in the course of the operation of the common market, one of the objectives of the Community and this Treaty has not provided the necessary powers, the Council shall, acting unanimously on a proposal from the Commission and after consulting the European Parliament, take the appropriate measures."
3. See for instance the Marchéal report in 1967. The Marchéal Committee was created in 1965 to "review national scientific and technological policies, reported to the first-ever meeting of the EEC Council of Science Ministers in 1967" (Peterson and Sharp 1998, 31).
4. See for instance Presidency Conclusions from Copenhagen in 1990, Dublin and Florence in 1996, Luxembourg in 1997, Cardiff and Vienna in 1998 and the so—called Vienna Strategy for Europe, Helsinki in 1999, and Lisbon in 2000.
5. EUCLID (European Cooperation for the Long Term in Defense) is a military equivalent to the civilian RTD program EUREKA (European Research Coordinating Agency).
6. This list is secret and has not been changed since 1958 (Mezzadri 2000).
7. During fall 1999 the Commission was reorganized and the General Directorates have changed names and to some extent tasks. The General Directorate on Industrial Affairs has been renamed the Enterprise Directorate. Furthermore, the directorates no longer use numbers to characterize them. However, in this chapter I use the old names as my empirical analysis focuses mainly on the period before fall 1999.
8. EDIG stands for European Defence Industrial Group, and AECMA stands for European Association of Aerospace Manufacturers.

9. One of my readers with insights into these matters questions whether DG III indeed has this strategy.

10. Already in 1993 when the advisory committee IRDAC (Industrial R&D Advisory Committee of the European Community) presented its position on the Fourth Framework Programme for research and development, it discussed the linkage between defense-related research and the framework program. "During the end of the Cold War period, the forthcoming years will probably show a rapid decrease in government expenditure on defence related research" (IRDAC 1993, 8). IRDAC is one of several advisory committees that have been set up to support the institutions of the European Union in the field of European R&D. IRDAC is the advisory body of the Commission in the field of science and technology comprising representatives from industry and other European associations such as research organizations, unions, and small and medium enterprises. In 1995 the Commission expanded the membership of IRDAC to include more "senior industrialists serving in a personal capacity (and appointed by the Commission), as well as five representatives of European 'peak-level' associations such as the European employers' federation (UNICE) and the European Trades Union Congress" (Sharp and Peterson 1998,178).

11. OJ C 167, 2.6.1997: 137.

12. This is also expressed by the Economic and Social Committee which argues that efforts must be stepped up to develop the possible synergistic effects between civilian and military activities, in both company strategies and public strategies (1997).

13. For an overview of European aerospace consolidation see, for instance, *Aviation Week&Space Technology*, 24 July 2000.

14. AECMA and EDIG are both based in Brussels.

15. Initially, the companies involved in the process toward a EADC were Aérospatiale, BAE, and DASA.

16. From 1985 it has been possible, according to EU company law, to create European economic interest groups, EEIGs (Pehrson 1999).

17. The statute will entail the creation of a European legal entity that will give the companies free movement within the single market. Companies will be able to merge across frontiers and to transfer their registered offices from one member state to another without any change in legal personality. Thus, there are organizational implications of creating a European company statute.

18. Interview with senior official DG V, September 1999. An important hurdle for the final decision on a company statute, which has been discussed for almost twenty-five years, concerns the worker involvement in the boards of the would-be European companies. Indeed, the issue of giving trade unions a vote in company boards is a controversial one. "If this hurdle can be overcome it could lead the way to the adoption of the Statute by the accelerated procedure foreseen in the Single Market Action Plan agreed at the Amsterdam European Council of the European Union" (COM (97), 466 Final). Another hurdle has been the old dispute

between the United Kingdom and Spain on the status of Gibraltar. This, in turn, is tied to several other issues within the European Union that have been the subject of dispute for several years. The dispute was, however, resolved in April 2000 (Chapman 2000).

19. SE stands for *societas europeas,* clarification by the author.

20. BAE stands for British Aerospace.

21. DASA stands for DaimlerChrysler Aerospace AG.

22. Airbus is a multinational European aerospace company.

23. Eurofighter is a joint multinational project for the creation of a common fighter aircraft among the United Kingdom, Germany, Spain, and Italy.

24. In the defense industry in Europe as a whole there are three giant companies: BAE Systems, EADS, and Thales (formerly Thomson-CSF) (Axelson and James 2000).

25. ESDI is an acronym that is used in a NATO context whereas ESDP is used within the EU.

Chapter 5
The Formation of a European
Organizational Field on Armaments

Thinking about military doctrines should be pursued in greater depth, as they are the framework to which industry refer. European states will in fact benefit from the economies of scale arising from the restructuring of their defence industries only if they procure common equipment. For European companies, restructuring will be attractive only if they have grounds for hoping that they will benefit commercially. For States, this restructuring will appear worthwhile if it holds out the prospect that they will be able to equip their armed forces with effective equipment produced at optimum cost.

—The Titley report,
European Parliament

OCCAR and the new WEAG Master Plan, with the final aim of a European Armaments Agency, the Letter of Intent (LoI) signed on the 6th July 1998, and the Action Plans of the European Commission must all be co-ordinated to support the strengthening of the European defence identity.

— EDIG Position Paper
February 1999

So far the empirical analysis has been focused on either the market field or the defense field. The purpose of analyzing the issue of armaments from this field dichotomy is to show how the issue is framed in two ways. The issue of armaments is thus argued to be either part of EU's defense process or an integral part of its RTD/industrial policy. The market frame dominated the early and mid-1990s, whereas the defense frame is strong in the late 1990s and at the beginning of the new millennium.

The next two chapters analyze how organizations that pursue different frames on the issue under study together form an organizational field. The empirical analysis shows that the field boundaries are rather porous and that the organizations from the two fields have begun to emphasize the need to handle the issue of armaments as a market and defense issue. This way of conceptualizing this issue is not new, but it became ever more salient during the 1990s and in the early years of this century as the legitimate way of framing the issue. Along with this gradual consolidation of a more complex frame on the issue of armaments, new forms of organizations and organizing around the defense industry and the issue of armaments were established. These new forms of organizations challenged the field dichotomy and the traditional domination and authoritative structures within the two fields. This formation of a European organizational field is studied focusing on three consolidating processes:

1. Reframing of the issue of armaments
2. New forms of formal/informal domination and authority structures
3. Scope and depth of organizational interaction.

Pillar One and a Half

In a statement issued by six European governments in 1998 it was declared that the "Ministers consider that a strong, competitive and efficient defence industry is a key element of European security and identity as well as of the European scientific and technological base . . . .Participation in the European armaments base should be balanced and should reflect the principle of interdependence" (Joint Statement 1998). This notion of the interplay between industrial strength and defense capacity was well stated by the French prime minister in a speech given in September 1998. "Our conviction is that the creation of an industrial and technological defense industry base, constituted by strong units, which primarily lean upon a market constituting the whole of Europe, is a necessary condition for the creation of a true common European defense industry" (Jospin 1998).[1]

In NATO documents, the economic side of the issue of armaments is also emphasized: "NATO should promote more cost–effective defence acquisition to meet evolving military requirements, thereby fostering a more efficient use by nations of limited national resources in research, development and production. Therefore, economic considerations should continue to play an important part in NATO's armaments activities" (NAC 1997, 7). It has been stated by NATO that the overall aims of the organization's armaments activities are economic, technological, and industrial. Indeed, a defense market is "necessarily primarily for security, but also for political and economic reasons" (Taft and Taylor 1992, 8). Furthermore, the work by NATO's Research and Technology Organization (RTO) seems to increase in importance (Andrén, interview, 1998). The market-oriented aspects of a closer European cooperation on armaments are also clearly stated in the work within WEAG and WEAO. One of the objectives of WEAG is "to open national defence markets to international competition" (see chapter 3). Furthermore, the existence of dual-use technology is also something that people within WEAG/WEAO and the EU are conscious of, for instance, the linkage between the EU's framework for research and development and the RTD activities within WEAO (see chapter 4).

Thus, organizations are aware that the issue of armaments has a dual character. The European Commission presented an early example of this awareness of the complexity of the issue in November 1997. In fact, in this communication the Commission presents a new way to conceptualize the issue of armaments and discusses how a more complex frame could be handled in practice. The communication (COM (97), 583 final)—"Implementing European Union Strategy on Defence-Related Industries"—was related to the earlier report on the defense industry (COM (96), 10). A memorandum with the same title had been presented earlier, in October 1997, by the two commissioners, Hans van den Broek (external relations) and Martin Bangemann (industrial affairs). The memorandum and the communication take a very pragmatic view of the defense industry and the market for armaments. The communication is very clear on the dual nature of the defense industry and does not pursue one perspective, but two—a community and a CFSP perspective. "An integrated European market for defence products must be set up using a combination of all the instruments at the Union's disposal: Community and Common Foreign and Security Policy, legislative and non-legislative instruments" (COM (96), 10, 2). The defense industry is both a "major means of production and essential to foreign and security policy. Any action by the EU has to take this dual nature into account, if necessary by adapting the resources within the Community's jurisdiction" (COM (96), 10, 5). The Commission suggests that Article 296 should be interpreted restrictively and that materials for the defense sector should be divided into three categories. This would mean that only sensitive goods—"highly sensitive goods" (such as nuclear)—would be covered by Article 296 (COM (96), 10).

The earlier tension between a market and a defense frame in the communication from January 1996 is clearly toned down. The defense issue can still be interpreted as part of a market frame, but the armaments market is a special market. DG III and DG IA, especially DG III, seem to have divided the issue in a more constructive work–sharing arrangement than was the case in the earlier communication from 1996 (Interviews, officials from DG IA and III 1998–1999). The new communication consists of two parts. The first part discusses a proposal for a common position on drawing up a European armaments policy with special emphasis on the creation of intracommunity transfers, public procurement, and common customs arrangements. The legal basis for this common proposition is Article J.2 in the Treaty on European Union (second pillar of the EU).[2]

The second part presents an action plan for defense-related industries, that is what the Commission considers being necessary measures "to ensure progress towards a true European market for defense products" (COM (97) 583, final, 5). The following actions are presented: intracommunity transfers, a European company statute, public procurement, RTD, standardization, customs duties, the innovation transfer of technology and small and medium-size enterprises, competition policy, exports dual–use goods and conventional armaments, structural funds, indirect taxation and direct taxation, principles for market access, benchmarking, and enlargement. The communication clearly entails a combination of the first-and second-pillar instruments. A European armaments policy would be linked to community policies (industry, trade, customs, the regions, competition, innovation, and research) and CFSP measures—it would be a "pillar one and a half."

The communication from November 1997 marked a reframing of the issue of armaments in a combined market and defense frame. This notion of the dual character of the armaments issue was not new. What was new was the fact that the Commission was so explicit in its recognition and that it presented measures on how to link these two pillar instruments and policy areas. In doing so, the Commission presented a proposal on a policy area in which it had very little formal competence. A timetable was presented in the communication on when various communications, Council decisions, etc., could be expected. The Commission was, for instance, preparing proposals for intracommunity transfers, a European company statute, competition policy, RTD, standardization, benchmarking, public procurement, and export controls in February 2000. However, this timetable has not been followed due to the stalemate in the Council, the anticipation of the LOI initiative (see below), and the resignation of the European Commission in March 1999. In February 2000 there were only proposals/reports on exports (code of conduct for arms exports adopted by the Council and a proposal from the Commission on new regulations on dual-use goods (COM (98) 258).

There have been many reactions to the communication on this dual character of the issue of armaments. The European Parliament stated as early as in the Titley report from 1997 (chapters 3 and 4) that there were both commercial and security aspects of the restructuring of the defense industry. This report has also, after modifications, been under consideration during 1998 and 1999 (A4-0482/98). In the revised report from 1998 the Parliament welcomes the action plan proposed by the Commission and it is seen as "a coherent set of measures aimed at extending the single market to armaments; therefore calls on the Commission to submit to the Council, on this basis, proposals designed to promote the gradual emergence of a common policy in the arms sector" (A4-0482/98, paragraph 6). The European Parliament recognizes the mutual interdependence between a European defense policy and a European defense industry that can match the American defense industry (A4-0482/98; Cars, interview, 1999; Carlsson, interview, 1999*).* "The European response to the challenge from trans-Atlantic industry is beginning to emerge, but needs to be vigorously supported by producer States and the Institutions and bodies which have responsibilities in this field in Europe" (A4-0482/98, paragraph 10). It is also stated that the proposed common position by the Commission "must be extended by common actions in the context of the CFSP aimed at setting up the structures needed for cooperation in the arms sphere, starting with the European arms agency, and at drawing up military programmes to meet the needs of the CFSP, in particular for carrying out 'Petersberg' missions" (A4-0482/98, paragraph 31).

EDIG supports the Commission's action plan. A memorandum from EDIG in February 1999 states that it is essential to establish a European armaments market (EDIG 1999). EDIG supports the EU policy on procurement and an export policy on armaments, but it is also recognized that the defense industry is a specific sector (EDIG 1999). The export issue is very important from the industrial perspective, but first it is necessary to establish the export control policy within the community (Woodcock, interview, 1999). Export control is perhaps the most sensitive issue in the action plan. The market for armaments must, according to EDIG, be created in accordance with a future common defense and foreign policy. "Thus, if the European defence industry is to survive to support the future common defence and foreign policy of the European governments with the appropriate high quality technologically advanced and price-competitive products, then a European domestic market of sufficient size must be established" (EDIG 1999). However, the governments must also listen to what the industry identifies as areas of excellence instead of determining what the industry should do (Woodcock, interview, 1999).

Furthermore, EDIG argues that the "R&T programmes for defence and civil needs are developed entirely separately, without any common synergies and in a totally uncoordinated manner" (EDIG 1999). "There are many technology areas, the so-called 'dual-use' areas, which are relevant to both defence and civil needs. In the future, the co-operative development of such technologies should be addressed by both national and international civil and military R&T authorities—for example the European Commission and WEAG—to avoid duplication. Industry should be part of this process through the participation of companies from both the defence and civil sectors, for instance, in the Framework Programmes of the European Union" (EDIG 1999).

The communication from November 1997 was under consideration within the Council of Ministers by March 2000, but it has not yet reached a common position. The day-to-day work has taken place within POLARM.[3] The disagreements in POLARM are multiple. Initially, there was a major concern over the political implications of a common position on armaments. A common position on this would mean an important step in forming a common defense policy, and the governments were reluctant to take such a defense decision in 1997 and 1998, that is, before the defense process in the EU was consolidated (chapter 3). The Commission's communication effectively took the opposite position; that is, that the Union should start dealing with practical issues (the fourteen actions) and not begin with the political decision on a common defense policy. This tension is, of course, less relevant after the political decisions in Amsterdam and Cologne (chapter 3).

The Commission's work on the action plan has not formally been dependent on a common position in the Council. A common position would, however, have given a clearer direction for the Commission in its effort to create an internal market for armaments. Indeed, it would have meant that the Commission would be recognized as an authoritative actor on the issue of armaments and that the Commission could take formal responsibility for this issue (Interview, Senior official DG IA 1999).

The reports from the POLARM meetings during 1997–98 on the Commission's communication in November 1997 show that there is a general resistance among the governments to give the Commission an important role on the issue of armaments. The cooperation should be inter-governmental and not supranational (POLARM, 1998d). It is also recogni-zed, however, that the issue concerns the first as well as the second pillars.

Already in 1996–97 a Franco-German proposal was presented in which the issue of armaments could be "an integrated cross-pillar system" (POLARM 1998d, 3). An overall aspect that is repeatedly mentioned in the reports from POLARM meetings is that it is important to take into account the special nature of the armaments sector when discussing the proposed action plan from the Commission. It is argued by POLARM that the fourteen community pillar instruments must be adjusted to this special policy area. This is the case with public procurement (POLARM 1998d). At present, community directives on public procurement do not apply where procurement is declared to be secret or where it is required for the protection of a member state's vital security interests. Opening national public defense procurement within the EU "should serve the immediate budgetary interests of the Member States because of the increased compe-tition" (POLARM 1998d, 3).

In Article 296 (formerly Article 223), it states, "Up to now, a number of Member States have made use of the possibilities provided by Article 223 not systematically to notify the Commission of concentrations of undertakings in the defence sector. In regard to State aid, Article 223 can-not generally be used, given the diversification of the activities of defence firms" (POLARM 1998d, 10). This position seems to lie near the view of the Commission, that is, to interpret Article 296 rather strictly (see also below).

The national positions are not always clear in the reports from POLARM meetings. Some governments are prominent by being absent (for instance the Swedish government) while other governments present rather extensive positions on the issue of armaments (for instance the French government). The French government is, in principle, in favor of the Commission communication and emphasizes the need for a joint strategy on European armaments policy (POLARM 1998c). The French government is also clear on the fact that the sector of armaments has unique features. Developing a European armaments policy "entails using CFSP instruments and that Community instruments may be brought into play" (POLARM 1998c, 4). This so-called balanced position is also taken by the Italian and German governments (POLARM, 1998h). It is, however, also clear that the governments, especially in France, favor an inter-governmental approach and that the French government is very reluctant to strengthen the role of the Commission in this policy area (POLARM, 1998h). A simplification of Article 296 is needed, but "as the CSFP stands at present, certain prerogatives of national sovereignty should be maintained for political reasons as regards export controls on complete systems or significant elements" (POLARM 1998d, 1). The French government also argues that Article 296 is needed in order to protect the European defense industry. Thus, an abolition of Article 296 would mean market access for U.S. companies (POLARM 1998d, 1). This position is also expressed as a general position of the POLARM group: "Unconditional access would dangerously expose the European DTIB to strong pressure exerted by U.S. industry which is highly competitive and enjoys major, destabilizing advantages" (POLARM 1998d, 13). The German government maintains that the Commission document's "comprehensive approach to the European defence equipment industry should form the basis of POLARM's discussions" (POLARM 1998g). The work by POLARM should, however, be limited to fewer topics than are presented in the communication from the Commission because POLARM lacks any mandate to deal with all of the fourteen topics in the action plan.[4] The British delegation considers that priority must be given to discussion of the issue of security of supply. It is stated that "industrial restructuring would be likely to increase the level of international dependency in all European nations. Governments would be required to accept that some domestic capabilities would be lost and procure directly from foreign or transnational companies . . . nations will need to be assured of supply and support of defence equipment between nations. . . . It will be essential for Governments to be confident of security of supply when entering into contracts with commercial companies, particularly with non-national companies, driven by economic demands" (POLARM 1998f, 1-2).

During fall 1998 and spring 1999 the activity within POLARM was very limited. It is also obvious that positions and papers from the major delegations in POLARM are being discussed within the new LoI framework (see below). However, in October 1998 the Commission presented to POLARM a nonpaper on procurement of armaments, outlining the principles that should direct procurement based on the first and second pillars. It is obvious that the issue of how to restructure the European defense industry has been a highly sensitive one in the Council of Ministers. In September 1999 POLARM presented a "Draft Common Position on Framing a European Armaments Policy—adopted on the basis of Article 15 of the Treaty on European Union." The document is suffused by caution and is not very rich in detail. However, the very fact that POLARM presents a draft on a common position is an important breakthrough in the political process on the issue of armaments.

In the preamble to the draft, the Cologne process is mentioned: "Whereas the European Council at its meeting in Cologne recognised the need to undertake sustained efforts to strengthen the European industrial and technological defence base and, accordingly, attached importance to making further progress in the harmonisation of military requirements and the planning and procurement of arms, as Member States consider appropriate" (POLARM 1999a, 1). In an annex to the draft proposal the specific characteristics of the armaments sector are outlined. Indeed, the two conceptualizations of the issue of the armaments—a market and a defense frame—are clearly recognized:

> There are a number of political and economic factors which have specific relevance for this sector. On the political side, there is the impact of the end of the Cold War and the changes that have resulted in the international climate in which foreign, security and defense policy are formulated and conducted. There is also the development of the EU's Common Foreign and Security Policy (CFSP). . . . On the economic side, the domestic side, the domestic demand for European defense equipment has been falling since 1987 and the global arms market has practically halved in the last decade . . . Employment in this sector has fallen by 37% since 1984. . . . This has hit certain regions . . . the market remains fragmented. . . . The lack of competition and impossibility of fully exploiting economies of scale has worsened the competition position of the European industry vis-à-vis the US since the 1980s. (POLARM 1999a, Annex to Annex 1, 1)

Furthermore, as in other POLARM documents, the defense-related industry is considered to be a strategic industry, which is not dictated by economic considerations alone. "Political, strategic and security considerations accordingly come into play in determining the conditions within which the industry operates and the demand for its products" (POLARM 1999a, Annex to Annex 1, 2). The emphasis is on the CFSP framework and thus on a defense frame on the issue of armaments. The Council of Ministers considers that a future common defense and security policy, supported by cooperation in the field of European armaments policy, falls within the framework of CFSP. It is linked to community policies, in particular on industry, trade, customs, the regions, competition, innovation and research. Its gradual development entails using CFSP or community instruments, as appropriate? Furthermore, a European armaments policy is "closely linked to the progressive framing of a Common Defence Policy" (Article 2).

The proposal for a common position from September 1999 has not been adopted by the Council of Ministers by December 2000. In fact, the work seemed to have come to a halt.[5] There are two main reasons for the political difficulty in reaching a common position in the Council of Ministers. The first reason concerns the CFSP aspects of the armaments issue. The problems within POLARM are thus linked to the general political process concerning a common foreign and security policy. However, the Cologne Declaration and the war in Kosovo have given this process momentum. There is a "spirit of Cologne," as one senior official in the Commission put it in September 1999 (chapter 3).[6] The Cologne Declaration and the ongoing process on the formation of a European defense policy should therefore have enhanced the possibilities for adopting a common position on a European armaments policy. However, in the late 1990s the issue seems to have left POLARM as the political center of gravity. Indeed, important political activities took place outside the EU within the so-called LOI initiative (see below). It can be argued that the LOI initiative between six governments in 1998 has had a clearly negative effect on the work within the traditional EU fora for the issue of armaments, especially from the Commission's point of view. It was not until that political process showed any results that the work within the Commission and the Council took a new turn.

The European Court of Justice and the Two Frames

The judicial aspects of the issue of armaments and the two frames are complex. This book does not deal with these legal issues in any detail, but it is nevertheless important to recognize the role of the Court in this study. Basically, the two frames on the issue of armaments activate two different legal frameworks, the one in the community pillar and the other in the intergovernmental framework of the second pillar of the European Union. In the defense frame, which was analyzed in chapter 3, the legal framework consisted of articles in the treaties and various articles in the CFSP framework. In chapter 4, in which the market frame was analyzed, the legal framework concerned different policy areas within the first pillar, articles on the internal market, industrial policy, RTD, and competition policies.

The tension between the two frames in this case has been subject to rulings of the European Court of Justice. In its ruling of 16 September 1999 (414/97), the Court argues that Articles 296 and 297 have "limited character . . . and do not lend themselves to a wide interpretation" (paragraph 21).[7] This ruling has been important for the Commission in its effort to establish a reinterpretation of Article 296 and to establish that the article "does not lend itself to a wide interpretation" (Carvalho 2000, 47). It has been argued that the ruling "shows how the derogation of Article 296 dealing with national security should be interpreted" (Carvalho 2000, 47).

In other related rulings on export control of dual-use goods (one of the actions in the communication from the Commission in November 1997) the Court has put commercial aspects, that is, the first pillar, before foreign and security aspects (the second pillar). In a ruling from 1995 the Court argued that "the Member States cannot treat national measures whose effect is to prevent or restrict the export of certain products as falling outside the scope of the common commercial policy on the ground that they have foreign and security objectives. . . . Consequently, while it is for Member States to adopt measures of foreign and security policy in their exercise of their national competence, those measures must nevertheless respect the provisions adopted by the Community in the field of the common commercial policy provided for by Article 113 of the Treaty" (C-124/95, see also C-70/94 and C-83/94).[8] A weak linkage between a measure and national security concerns has not been regarded as sufficient. The regulation on export of dual-use goods is thus based on two legal instruments: Article 113 in the TEU and joint action under the Common Foreign and Security Policy (J3)—which together form an integrated system (*Official Journal* No 367, 31/12 1994).[9] Dual-use goods are defined as "goods which can be used for both civil and military purposes" (*Official Journal*, Article 2). However, the Commission in 1998 presented a proposal for a new regulation regarding the export of dual-use goods, which will only be based on the community pillar (COM (98) 257).

Reframing or Two Frames?

Interestingly, the frames—market and defense—are not necessarily linked to specific organizations. They move across organizations that are oriented toward the EU's civilian power and organizations that are dealing with security and military affairs. NATO pursues the market frame, and the defense frame has been part of the Commission's work on the issue of armaments. One reason for this frame mobility is that the organizations seem to change how they define their roles and tasks in European politics. In the wake of the end of the Cold War, NATO is becoming more oriented toward economic, technological, humanitarian, and other more civilian-oriented issues (chapter 4), and the EU (including the European Commission) is becoming more involved in defense and military affairs (chapter 3).

The communication from the Commission (COM (97), 583 final) shows that there is a third way of framing the defense industry and equipment issue, namely, a modified market perspective. The earlier tension within the Commission suggests that frame competition takes place not only between organizations but also within individual organizations. It can be argued that various parts of the European Commission are part of both the market and defense fields. Thus, the Commission's relationship with the external environment is complex and multifaceted. The new communication from the Commission, in which it is argued that the armaments market is a special market, once again underlines that framing is not a static activity. The process of frame competition not only promotes various interests in the Commission, but also results in *re*framing (Mörth 2000a). Frame competition is part of the communication and interactions within the organization. This common frame is important to build a capacity within the armaments and industry issue area. This does not mean that there are no more conflicting frames, but by *re*framing the armaments and industry issue the Commission manages the conflict between the market and defense frames. Thus, frame competition will persist, but in a less obvious way. The trick is to maintain some ambiguity while at the same time being able to move forward in the policy-making process (Sahlin-Andersson, 1998). The sense-making process, in which various frames are presented, functions as an important component in generating cohesion within the Commission. Various parts of the Commission have achieved some common understanding of how to proceed in the policy-making process, while at the same time maintaining a basic lack of clarity concerning how these two frames could be reconciled.

It is obvious that POLARM and other intergovernmental fora recognize the mutual interdependence between a European defense policy and a European defense industry. However, the POLARM documents suggest that the recognition of the mutual interdependence between the two integration processes does not necessarily mean that the issue of the armaments has been reframed. The issue is still framed as an issue within the second pillar of the European Union. The work within POLARM appears to be aimed toward changes within the current field structures rather than toward any reframing such as that attempted by the European Commission. By framing the issue as belonging to the defense field, the member governments can keep the Commission and the rules from the first pillar outside the discussion. The rulings by the European Court of Justice also suggest that the two frames still exist while at the same time underlining the tensions between the market and a defense frame.

Thus, a general conclusion is that there is growing recognition of the interdependence between the two frames. This recognition has been explicitly pronounced not only by the European Commission but also by the member governments of the EU. It is, however, somewhat unclear whether this recognition also entails a new thinking, especially concerning POLARM and EU governments, or if the actors involved still proceed from two organizational fields and modes of thinking about the issue of armaments. What is clear is that the organizations from the two organizational fields have realized that they are dependent on each other. They are also struggling over something they share: the problem of how to define, categorize, and interpret the issue of the armaments.

New Forms of Formal/Informal Domination and Authority Structures

The previous chapters have analyzed how the traditional organizations, the EU, WEAG, and NATO, have discussed and handled the issue of armaments. Important organizing activities have also taken place outside these organizations. These new organizational activities are dif-

ficult to classify in terms of the two frames—market and defense. They
cross the traditional organizational field boundaries and emphasize the
dual character of the issue of armaments. In contrast to the Commission
they are not convinced that the best way to cooperate and organize on this
issue lies in the European Union. They thus challenge the traditional organizations and their efforts to control and frame the issue of armaments.

OCCAR

In November 1996 the Joint Armaments Cooperation Organization,
OCCAR (Organisme conjoint de coopération en matière d'armement)
was created to act as a joint program office on behalf of France, Germany,
the United Kingdom, and Italy. The decision to create a new cooperative arrangement was already announced in Baden-Baden in 1995 by the
French and German defense ministers; in 1996 it was named OCCAR and
the cooperation was extended to include Italy and the United Kingdom
(*L'Armement*, March 1998).[10] Five principles for the cooperation were
decided in Baden-Baden that were in extenso reiterated in 1996 (OCCAR
Strasbourg 1996, see also Prévot, 1998):

1. Preeminence of the criterion cost/efficiency in the choice of indu-
    strial goods
2. Harmonization of operational needs and technology policies
3   Strengthening of the European industrial base
4. Abandonment of juste retour in favor of a principle of a more
    global return
5. Open to other countries.

The document hardly ever explicitly mentions defense equipment (or
armaments) although the cooperation is signed by defense ministers and
has the aim of launching programs that are defense oriented (OCCAR
1996).[11] There is thus a clear emphasis on industrial goods and the European industrial base. Furthermore, the new cooperative arrangement
does not encompass either the principle of juste retour or research and
development as is the case within WEAG and WEAO (chapter 3). The
traditional principle of juste retour has been replaced by a more global
principle of return. This new principle has been interpreted as a political
determination to create cooperation based on more commercial principles

and with a long-term perspective, which is contrasted with the cooperation principles within WEAG (Mezzadri 2000). The difference between the juste retour principles in OCCAR and WEAG has been described in terms of economic efficiency versus a social security approach to procurement of armaments (Heisbourg 2000). Indeed, the creation of OCCAR is seen as something other than the traditional organization of WEAG. The new features are the emphasis on competition, commercial principles, and the principle of global, rather than national, *juste retour*.[12] OCCAR has also been regarded as a flexible cooperative arrangement. The organizational setup is loose, and until 2001 it had no legal personality. The system of exchange of information between the participants and the close interaction among the personnel from the four participating countries suggest that the cooperation has been a far more coherent and homogeneous arrangement than seems to be the case within WEAG (Prévôt 1998; CEPS 1999; Mezzadri 2000).

As mentioned in chapter 3, the WEAG/WEAO framework has been discussed as the beginning of a European armaments agency. Originally OCCAR was meant to be a subsidiary agency of the WEU framework (de Briganti 1997a). In fact, when WEAO was established it would have provided the institutional cover for the Franco-German bilateral armaments agency that was expected to evolve into a multilateral agency (de Briganti 1996). The incorporation of OCCAR into the framework of the WEU was supported by the intergovernmental agreement in 1996 by the four participating countries in OCCAR. In the agreement it was stated that "ayant pour objectif l'etablissement dans les meilleurs delais d'une organisation ayant la personalité juridique, en tant qu'organe subsidiaire de L'Union de l'Europe Occidentale (UEO)" (OCCAR 1996). The European Parliament also expressed the wish for a merger between OCCAR and WEAG "to form the European Armaments Agency referred to in the Maastricht Treaty" (A4-0482/98, paragraph 10; see also below). Indeed, the WEU route had for several years been regarded as the most promising one for OCCAR. This would give the WEU a role to play in an important policy area and would resolve the problem of OCCAR's legal status (de Briganti 1997b).

However, OCCAR has not been incorporated into the WEU. One reason for this is a resistance within the WEU, especially from the smaller countries that lack a defense industry (de Briganti 1997a; de Briganti 1997b; Hitchens and Tigner 1998). The four countries within OCCAR represent approximately three-quarters of the EU's defense expenditure and 80 percent of the EU's procurement spending (Heisbourg 2000). The countries that have no defense industry, or only a weak one, have feared that incorporation of OCCAR into the WEU will lead to closed projects in which only the countries with strong defense industries will be able to participate. There has also been resistance from the OCCAR countries to being incorporated into what is perceived as a rigid organization.[13] Their impatience with the slow progress of WEAG appears to be an important reason why the French and German governments announced the new initiative in 1995 (Prévôt 1998; Schmitt 2000a). Another reason for having OCCAR as an independent organization is national autonomy (Britz 2000).

In September 1998 the legal status of OCCAR was enhanced and the link to WEU weakened (OCCAR 1998, see also Mezzadri 2000). In January 2001 OCCAR gained its own legal personality. A legal personality entails that all programs offices (there are now four) will "apply common regulations and management procedures" (Schmitt 2000a, 3). OCCAR will also be able to accept multiyear agreements from the governments, and on their behalf enter into contracts with industrialists for various programs. Paradoxically, the flexibility that has been the trademark of OCCAR, with a loose organizational setup and a system of information exchange between the participants, can thus be transformed into an agency with a more formal decision-making structure in which the principle of global *juste retour* can be difficult to uphold. However, by strengthening its legal personality, OCCAR can present itself as a more forceful organization and a legitimate actor in the process of organizing European cooperation on armaments.

The Letter of Intent (LOI) and the Framework Agreement

A letter of intent was signed in July 1998 between the defense ministers of the countries that participate in discussions regarding the future European aerospace company: France, Germany, the United Kingdom, Italy, Spain, and Sweden (chapter 4). In July 2000 a general framework agreement was signed by the six governments. This was an important political initiative since the six countries represent more than 90 percent of the EU's defense industrial capacity (Heisbourg 2000). Already in December 1997 the three governments in France, Germany, and the United Kingdom declared that it was important to strengthen the European defense industry, especially the aerospace sector. The analysis below focuses first on the initial steps in the LOI process and moves on to an analysis of the framework agreement.

The political initiative by the three governments in December 1997 and the succeeding LOI have been discussed as a reaction toward the demands and pressure from the European defense industry, especially from the aerospace industry (chapter 3). Indeed, the LOI process started as a helping hand to the industry, but has evolved into a more independent and comprehensive political process. It also started as a political initiative to support the aerospace industry but has gradually encompassed the European defense industry in general. Furthermore, although the United States is hardly mentioned in the political statements, it is obvious that the Other is omnipresent and that the American consolidation of its defense industry was an important driving force for the European political process (von Sydow, interview, 2000).

The overall ambition with the LOI was to enhance the conditions for the defense companies, that is, through market-making activities. "The Participants desire to establish a co-operative framework to facilitate the restructuring of European defence industry" (Letter of Intent 1998). "European defence companies are already actively engaged in a restructuring process. To facilitate this process, and to take advantage of the opportunities it offers, while preserving national defence capabilities, timely government action is necessary" (Report to the Ministers from the Executive Committee, 1999). As French Defense Minister Alain Richard put it in a speech on 1 July 2000, "[The European defence industry] won't reach its potential unless European governments decide to give an impulse to industry research and to acquire the armaments in a harmonised manner, in a fashion that is directed toward future developments."[14]

Initially the LOI did not represent any legally binding commitment, and an important glue that would hold the countries together was instead the notion of mutual interdependence. However, the notion of mutual interdependence is not developed in the documents issued by the six ministers. The Swedish government, one of the participants in the LOI, has in other documents stressed how the end of the Cold War has opened possibilities for cooperation based on mutual interdependence (Mörth 2000b). European collaboration is essential in order to maintain national technological capacity. "Our future industrial capacity will have a structure where mutual industrial dependencies across national borders will be an important factor" (Swedish Government Bill 1998/99:74, 116). This means that the Swedish defense industry must be part of the ongoing process of internationalization and Europeanization. The Swedish government recognizes officially that national self-sufficiency is no longer realistic. The future for the problem of security of supply in a crisis (or war) must therefore rest on mutual interdependence. Establishing mutual interdependence means that "a supplier of some critical component or subsystem depends on deliveries from your company, or at least a company based in the same country" (Swedish Government Bill 1998/99, 74, 35-36). The British government has also recognized that consolidation of the defense industry "may involve the loss of some domestic industrial capacity in order to preserve other capabilities. This is leading to more mutual interdependence between nations and companies alike. Governments need an assurance of security of supply, just as companies need to know the procurement plans of governments to construct viable business structure" (Ministry of Defence 1999, paragraph 99). The LOI initiative is therefore welcomed by the British governments (Ministry of Defence 1999).

It is obvious that cooperation based on mutual dependency means that there would be centers of excellence with a single source structure. The Swedish and other governments must thus identify those strategic competencies that must be kept in the country and those strategic technologies that can be produced elsewhere. In the LOI, the six governments have stated that accepting mutual interdependence is to accept "abandoning" industrial capacity and that they would not be allowed to have direct influence on a transnational defense company (TDC) (Letter of Intent 1998). These companies will be run on a commercial basis, have private capital markets, and be listed on the stock exchange.

The organizational setup within the LOI consists of an executive committee, which is composed of high officials from each country. There have also been six working groups (Security of Supply, Export Procedures, Security of Information, Technical Information, RTD, Harmonization of Military Requirements) consisting of officials from the six governments.[15] The working groups have been responsible for providing "policy advice to, or undertaking specific tasks for, the Executive Committee" (Letter of Intent 1998, 8). Thus, in the early LOI process the governments identified six major areas in which there were perceived obstacles to industrial restructuring at the European level. Reports from the LOI working groups and the executive committee were presented in July 1999. In the following section the various reports are discussed, and the section concludes with a general analysis of how these reports can be seen in relation to the activities within the EU, the WEU, NATO, and OCCAR. The following questions have been put to the reports: How are armaments framed, that is the thinking of the issue of armaments, and what are the recommendations on what has to be done? How is the LOI framework discussed in relation to other organizations that deal with the issue of armaments?

The Working Group on Security of Supply addresses one of the crucial dilemmas in the ongoing restructuring of the national defense industry, namely, the national need to regulate the prioritization of supply in time of an emergency, crisis, or war. Supply in time of peace will be negotiated under "normal commercial practices" (Report of the Working Group on Security of Supply, 2). The report states that the participants will consult each other in an emergency, crisis, or war "in order to enable the requesting Participant(s) to receive priority in ordering, reallocating or modifying Defence Articles and Defence Services" (2). This supply problem is especially complex in situations where there are transnational defense companies, which weaken the role of the state as a gatekeeper and regulator. This is not considered to be a major problem according to the working group, but the group concludes this section of the report with the statement that "it may be necessary in some cases to have restrictions on foreign ownership or control of a TDC for reasons of national security"[16] (Report of the Working Group on Security of Supply, 9). Another problem is the fact that there are different national legal regulatory frameworks and that these differences can present an obstacle "to the establishment of binding measures in certain areas such as control of foreign investment in a TDC" (Report of the Working Group on Security of Supply 1999, 2).

The group's analysis of the possibility of overcoming the problems of the national requirements for the timely supply of armaments is the overall idea of the LOI initiative—mutual interdependence. "The Security of Supply Working group has drawn up a series of measures to ensure the continuity of supply for the LoI nations. These are based on early and effective consultation to secure the interests of the six nations" (Report of the Working Group on Security of Supply, 6). The group stresses the need for a consultation process between the six governments and that the industry in the countries inform their governments "in advance of their intention to form a TDC or significantly change its composition or strategic defense activities. As soon as a Participant becomes aware of such an intention the Participant will inform the other involved Participants" (Report of the Working Group on Security of Supply, 1999, 4). The Security of Supply working group seems to continue the work that already started within POLARM, especially by the British delegation. In the POLARM paper the British delegation clearly points at the problems with private and transnational companies and indicates that closer intergovernmental cooperation, based on voluntary commitments between the participants, is an important way to secure supply in a crisis situation (POLARM 1998a; POLARM 1998c). Interestingly enough the work in POLARM on the issue of security of supply is not mentioned in the working group within the LOI framework.

Harmonization of military requirements and the formation of common military capabilities are discussed in the Working Group on Harmonization of Military Requirements. The group links the perceived need to harmonize military requirements with the new security environment in the wake of the end of the Cold War. "Developed societies and their armed forces depend increasingly on the undisrupted use of information and communication networks. While advances in technology bring new opportunities and provide enhanced capability they also introduce new risks and vulnerabilities" (Report of the Working Group on Harmonization of Military Requirements, paragraph 304). These new types of security challenges mean that cooperative and "collective approaches are required" (Report of the Working Group on Harmonization of Military Requirements, paragraph 306). The armed forces in Europe must be "flexible," "mobile," "deployable," and "sustainable" (paragraph 506). Similar to the CNAD reports (chapter 3), the group emphasizes the need for interoperability with allies or coalition partners (paragraph 511). The report presents a concrete analysis of military equipment areas and identifies the military equipment area in which harmonization has the greatest potential (the information and intelligence area). The group concludes that territorial defense, maritime protection, land protection, etc., are decreasing in importance, whereas information, intelligence, deployability, mobility, force protection, etc., are increasing in importance.

In a way that distinguishes it from several of the other working groups, the Working Group on Harmonization of Military Requirements explicitly discusses other organizations that are dealing with defense issues.[17] It is argued in the report that "the responsibility for formulating military requirements and the link to the defence planning process in NATO are now agreed in principle" (Report of the Working Group on Harmonization of Military Requirements, 1999, 20). Military requirements "of the NATO nations are harmonised within the Alliance as part of the collective defence planning process" (Report of the Working Group on Harmonization of Military Requirements, 1999, 25). In the European fora there are Finabel (that is, closer military cooperation between NATO members and the standardization of land-based equipment), Eurolongterm (which deals with the development of operational doctrines and related user requirements for armaments), and WEAG (which concentrates on armaments acquisition cooperation within Europe). The Working Group on Harmonization of Military Requirements argues that the aspect of harmonization of programs and equipment timetables has been neglected. Panel I in WEAG suffers from major shortages, in the view of the group. "There has been no lack of will or effort to co-operate on harmonising military requirements. However, the principal weakness has been a lack of co-ordination and interlinkage of the overall process of harmonising military requirements within Europe" (Report of the Working Group on Harmonization of Military Requirements, 1999, 21). "Within almost all fora, developments in the recent past have taken the focus from requirements harmonisation to focus almost entirely on investigating possibilities for equipment co-operation" (Report of the Working Group on Harmonization of Military Requirements, 1999, 23).

The Working Group on Harmonization of Military Requirements concluded that it is important to "facilitate the emergence of a homogeneous demand towards industry" (Report of the Working Group on the Harmonization of Military Requirements, 1999, 23). To achieve a common demand on the industry, the governments need to cooperate as early as possible in the "genesis of the requirement and maintain this cooperation throughout the process up to the specifications of the systems they want to develop and/or purchase."[18] Clearly, this is a complex harmonization process and it includes harmonization of equipment acquisition procedures and the formulation of defense policy through a common conceptual analysis. Thus, harmonization of military requirements has implications for defense planning. Two initiatives are mentioned in the report, namely, DCI—the Defense Capabilities Initiative launched during the Washington NATO summit in April 1999—and the initiative in October 1998 in which an offical at WEU was tasked "to provide a consolidated status report on current approaches to identify military operational requirements which contribute to force and armaments planning and to develop proposals" (Report of the Working Group on Harmonization of Military Requirements, 1999, 22). Furthermore, the group argues that in "national planning the process of deriving a military requirement ideally starts as early as the national security and defence policy is being formulated. Therefore, logically, internationally the harmonisation process should also start at this stage" (Report of the Working Group on Harmonization of Military Requirements, 1999). However, this requires a common security and defense policy. Thus, "in the absence of an ESDI the next opportunity for harmonisation occurs with the comparison of military concepts and doctrine" (Report of the Working Group on Harmonization of Military Requirements, 1999). This perceived lack of a European security and defense identity was less relevant in 2000 when the European Union was in the process of building such an identity (chapter 3).

Similar to the Working Group on Harmonization of Military Requirements, the Working Group on Research and Technology criticizes existing European cooperation arrangements. The group recognizes that special organizational arrangements are needed but that these arrangements also "have to be set up in an incremental process taking into account present fora and their respective development, and agree that creating a new organisation or forum is the last option" (Report of the Working Group on Research and Technology 1999, 2). A first step is to discuss with WEAG/WEAO "their ability to adapt their working patterns in order to meet all the Requirements" (Report of the Working Group on Research and Technology 1999, 3). One important requirement is the possibility of so-called closed projects, that is, those research and technological projects in which only a limited number of countries can take part. This could be a controversial issue within WEAG/WEAO since these bodies do not exclude any member country from participation in various projects (chapter 3). The Working Group on Research and Technology is in favor of the general principle of flexible integration, and the report concludes that a crucial issue that lies ahead concerns WEAG/WEAO's willingness and ability to adapt to the requirements of the LOI initiative. An important issue in the report on Research and Technology is therefore the future relationship between the LOI initiative and WEAG/WEAO, for instance, the legal framework of LOI. Should LOI build a new legal personality or should it lean on the existing fora? The group states that a memorandum of understanding would allow any combination of the LOI signatories to conduct research and technology programs, including technology demonstrators, with transnational defense companies (Report of the Working Group on Research and Technology, 1999, 12). Thus, it would give the LOI countries flexibility. The relationship with TDCs will be guided by a special code of conduct. Furthermore, the group argues that the ministers should agree on the placing of contracts by an "international body acting as the agent of the LOI six" (Report of the Working Group on Research and Technology 1999, 13). It is also argued that there are advantages in WEAO since five of the six LOI members have the appropriate legal relationship with the WEU for placing contracts, "and an established mechanism exists to create such a relationship for the sixth (Sweden)" (Report of the Working Group on Research and Technology, 1999, 13). The WEAO Research Cell is also the precursor to a full European armaments agency, "which could be expected to supersede and 'sweep up' existing fora such as the LoI six in due course" (Report of the Working Group on Research and Technology, 1999). Thus, the group does not exclude the possibility of tying the legal framework of LOI to the existing fora. The working group only considers a new legal personality a fallback position.

The Working Group on Security of Information addresses the dilemma between national industrial security provisions and the activities of a transnational company (movements of personnel, information, or materiel). Indeed, security classifications differ among the six countries (Report of the Working Group on Security of Information, 1999, 4). In contrast to the other reports, this one shows that there are national disagreements on how to go about cooperating on information security: "The main issue in dispute is that whilst Germany considers that any nation belonging to the European Union who holds an appropriate Security Clearance is allowed to have access to Classified Information without the prior consent of the originator country signatory to the LOI, the rest of the countries consider that such a rule only provides for the nationals of the countries signatories to the LOI" (5). There is also a dispute over access to classified information for individuals holding a dual nationality. Basically, the disputes concern the interpretation of international law, that is, the ways in which the LOI countries can decide on how nationals from non-LOI countries can obtain classified information (for instance, the Vienna Convention on the Law of Treaties between States and International Organizations from March 1986). A crucial issue seems to be how to handle the United States in this matter. A general issue that is addressed is thus how open the LOI initiative and the future activities should be toward other European countries and the United States. Is LOI an exclusive club for just six countries, or is it an important building block in a more comprehensive European effort to enhance industrial/technological capacity?

A similar dilemma between the perceived need to cooperate and the fear of losing national control is discussed in the Working Group on the Treatment of Technical Information ("recorded or documented information of a scientific or technical nature whatever the format, documentary characteristics or medium of presentation") (Report of the Working Group on the Treatment of Technical Information 1999, 4). This dilemma is, for instance, expressed in the following way: "There may be a requirement to restore a national capability for Defence Articles of Defence Services in exceptional circumstances by means of commercial licenses, which if required, will include Leader/Follower provisions" (Report of the Working Group on the Treatment of Technical Information, 1999, 8). By "leader" the report means "a TDC which furnishes the necessary technical assistance and technical information to a follower to enable that follower to become a source of national development, manufacture or repair for a Participant" (Report of the Working Group on the Treatment of Technical Information 1999, 4).

One of the politically sensitive issues in the LOI work is the simplification of intra-LOI transfers (armaments exports). Thus, the issue of export control is discussed only within this limited cooperation arrangement. One motive behind a simplification of intra-LOI transfers is, according to the Working Group on Armaments Exports to minimize bureaucratic rules and to give the national and transnational industry a "stable basis for future planning" (Report of the Working Group on Armaments Exports 1999, paragraph 2). The licensing procedures have to be simplified, and the group presents new procedures for transfers of defense articles between the LOI nations. "This would be achieved by the adoption of open export licenses that would permit multiple shipments of defence articles" (Report of the Working Group on Armaments Exports 1999, annex 4). Approval for export of defense articles outside the LOI group "would be reached by consensus on a project by project basis" (Report of the Working Group on Armaments Exports 1999). The final agreement to grant an export license, however, would lie "with the government where final assembly took place or from where the export would be executed" (Report of the Working Group on Armaments Exports 1999). These measures should "be consistent with the right of individual LOI partners to continue to exercise sovereign decisions on defence exports in accordance with their national foreign and security policy in conformity with article 223 of the Treaty of Rome" (Report of the Working Group on Armaments Exports 1999, 2). The report thus avoids the crucial issue, namely the question of whether a simplification of intra-LOI transfers would not also put pressures on the harmonization of national export rules. However, the group recognizes that "simplification of intra-LOI transfers and harmonisation of national export control policies are closely linked" (Report of Working Group on Armaments Exports 1999, paragraph 3).

To sum up, the reports are rich in detail and they also contextualize the issues in a broader political, security, and technological perspective. They cover civilian aspects of the defense industry as well as more defense-and security-related aspects. Positions taken by the individual governments are mentioned in only some of the reports. In general, the working groups discuss the issues as a collective.

The six reports have been discussed within each country and functioned as an important phase in the political process, which in July 2000 resulted in a general framework agreement that was signed by the six governments—the so-called Farnborough agreement. The national parliaments will ratify the agreement, which indicates that it has a rather strong legal status. Time-consuming national ratification can be avoided since two countries can proceed with their cooperation as they have ratified the agreement (Article 55 in the framework agreement). So, in contrast to the LOI in July 1998, the framework agreement is a legally binding document. Until the conclusion of the framework agreement some of my interviewees suggested that one possibility was to decide on a general agreement—a chapeau agreement—which was to be followed by varying types of intergovernmental agreements. This seems to be the path that the six governments have taken.

The agreement from July 2000 reinforces and confirms the earlier joint statements and working group reports of LOI. On some points the agreement is more clear and elaborated than in earlier documents. The executive committee is given a more permanent status. It will exercise "executive-level oversight of this Agreement, monitoring its effectiveness, and providing an annual status report to the Parties" (Framework Agreement 2000, Article 3). The committee shall meet "as frequently as necessary for the efficient fulfilment of its responsibilities" (Framework Agreement 2000). The articles on the six cooperation areas confirm what has already been stated in the working group reports. One novelty, in comparison with the working group reports, is that a global return is desirable in common defense-related research and technology activities (cf. OCCAR). Another new feature is that the area of information security has been supplemented by a new issue/category: the protection of commercially sensitive information. Furthermore, the importance of ensuring the parties' security of supply is more emphasized than in earlier joint statements. It is also more clearly stated than in the earlier reports that the participating governments must develop "harmonised force development and equipment acquisition planning" due to the emergence of transnational companies and reduced or abandoned national defense industrial capacity (Framework Agreement 2000, Article 45, a).

In contrast to the Working Group on Armaments Exports, the framework agreement explicitly mentions the EU's code of conduct for arms in connection to the issue of transfer and export procedures with countries outside the LOI circle.[19] This has been interpreted as the important legitimation base for the entire LOI agreement (von Sydow, interview, 2000). The six governments thereby guarantee that they will not pursue a different export policy from that which is decided within the EU. The objective of the agreement is to "bring closer, simplify and reduce, where appropriate, national export control procedures for Transfers and Exports of Military goods and technologies" (Article 1). Two ways of achieving this are proposed, namely, to simplify and reduce export control procedures for transfers of defense components among the six signatory countries and by collectively agreeing in advance on a list of permitted export destinations to countries outside the LOI circle (Articles 12 and 13).

The new features and clarifications in the framework agreement raise the question of what kind of specific agreements will follow from the general agreement. Legally binding agreements could be politically sensitive since they would entail perceived losses of national sovereignty. Less binding agreements are also problematic since they would be easier for countries to break, for instance, in the case of security of supply. The LoI governments have argued all along that the cooperation must be based on consultation and communication, that is, on mutual interdependence and trust. The rationale behind this notion is the fact that it is not possible to handle the problems the states are facing nationally in a situation where the defense industry is deregulated and the national regulatory framework is loosened. It is, of course, easier to obtain trust in a circle of six governments that pursue a relatively homogeneous policy than it is with the fifteen or even more governments, with their greater variety of defense industry structures and defense industry policies, that may be envisioned in the future.

As already mentioned, the notion of mutual interdependence as the new glue between the countries is not developed in the individual reports or in the common reports by the six governments. In the agreement, mutual dependence is sometimes referred to instead of mutual interdependence. It is difficult to see whether this means anything in substance or if the terms are used interchangeably. It is obvious that the LOI initiative—the documents from the ministers and working groups—tries to handle a tension between the two frames on armaments. The end of the Cold War has put the European governments in a dilemma—a dilemma that placed them between an emphasis on national security interests, on the one hand, and the internationalization of economy and technology, on the other. A closer relationship between civilian and defense-related industry is needed for reasons of economic competitiveness, but a strong European defense industry is also an important foundation for a European defense identity and capacity. Now, when it is time to move from talking about mutual interdependence to actually implementing it, the whole notion will be tested.

An important issue in the work done by these six working groups concerned how the LOI initiative should relate to the already established organizational and institutional arrangements such as OCCAR, NATO, WEAG/WEAO, POLARM and the work on the action plan within the European Union. In some of the reports, this issue is explicitly discussed (for instance, in the research and technology report), while in other reports, like the export control report, linkages between the LOI initiative and the work within the EU are less problematized. Although the groups seem to be aware of the need to link the LOI initiative to the existing organizations, they criticize the current organizations, their organizational setup, decision-making structure, and their activities. Indeed, the LOI initiative challenges the traditional way of dealing with the issue of armaments, especially concerning what measures are necessary. The issue of armaments belongs to both the market and defense frames. This is in line with the communications from the Commission. However, it is argued by the Research and Technology working group that the issue of armaments has to be handled differently from in the WEAG/WEAO framework. This framework is perceived as more or less obsolete, at least according to the Research and Technology working group. The activity in NATO is seldom mentioned in the LOI reports. It is, however, stated in one of the reports that the NATO armaments activities must adapt to cooperation activities that take place outside the alliance.

Paradoxically, the action plan from the Commission is never mentioned in the reports, although several issues that have been discussed in the six working groups are similar to the issues discussed in the Commission's action plan in November 1997, for instance, issues concerning intracommunity transfer, export, research, and technology. This can be explained by the fact that the people who have been involved in the LOI process mainly come from the defense ministries in the six countries. The organizational environment is therefore organizations from the defense field and not the European Commission within the market field. So, even though it is clear that the overall LOI initiative recognizes the market aspects of the issue of armaments, these aspects are seldom directly addressed in the reports. Furthermore, the LOI working groups do not mention POLARM, which is a working group within the Council.[20] This is rather strange since several issues that are addressed in LOI have also been discussed in POLARM and sometimes in a rather similar way—and sometimes by the same people. The question of security of supply is not only important in the LOI framework but is also at the center of the work by POLARM (see above). Thus, discussions on the issue of armaments seem to move among different cooperative arrangements.

The lack of reference to other organizations cannot be interpreted as ignorance on the part of the officials and politicians in the LOI process. During the work within LOI there was clearly a political awareness that there are multiple processes and actors on the issue of armaments (von Sydow, interview, 2000). It is therefore plausible that the texts' exclusion of the European Commission (a supranational organization) and its activities on the issue of armaments was a clear statement from the governments—this is an issue for the six governments only and it primarily belongs to the intergovernmental path of the European integration process. It is also plausible that, by excluding any reference to POLARM, the LOI participants made it clear that LOI is something apart from the European Union—that the LOI is an actor in its own right. The relationship between the EU and LOI seems to have been rather sensitive during the early phases of the work by LOI. Gradually, however, an incorporation of LOI into the EU seems to be a less controversial issue, although the political view among the six governments is that it is too early to communitarize LOI (von Sydow, interview, 2000). First of all, the agreement must work; projects must be launched until the governments can discuss the linkage to the EU (von Sydow, interview, 2000).

A European Armaments Agency (EAA)

A would-be European armaments agency is mentioned in the Amsterdam Treaty (chapter 3), but apart from this general political statement there are no concrete decisions on how to create such an agency. In fact, the question of creating a European armaments agency goes back to the general and politically controversial question on the decision-making structure of such an agency (supranational or intergovernmental) and how comprehensive the agency should be, that is, if it should comprise all the EU member states, the participants in WEAG and WEAO, or if its core should comprise only the countries in OCCAR and LOI.

In the masterplan from November 1998, written by the chairmanship of WEAG with the assistance of a group of national experts, it is argued that a European armaments agency could be created that is independent of the EU, even though the rules within the community pillar must be recognized. A fully fledged agency requires "a common armament market with harmonised rules in order to guarantee equal competition conditions for suppliers in all WEAG/WEAO countries" (Masterplan, 1998, 8). However, it is unclear in the report what kind of relationship this agency will have to the EU and what kind of decision-making authority will be given to the would-be agency. The report seems to support the formation of two emerging legal frameworks—the procurement rules within the community and procurement rules within the agency.

The two organizations, WEAG and WEAO, have been active in pursuing a policy in which both will have a central role in a deepening European cooperation on armaments. The work on the future EAA continues and is based on the masterplan agreed on by the WEAG ministers on 12 November 1998 at their meeting in Rome. However, the political support for this work had clearly diminished. It was regarded as something no longer needed because the WEU was going to be incorporated into the EU. In a new version of the masterplan in autumn 1999, compatibility, and not competition, is presented as the key word in the organizational mosaic: OCCAR and the legal framework of the EU would thus be part of this new umbrella organization (Schlieper, interview, 1999). A "holding" (an umbrella grouping) of existing organizations is the solution (Schlieper, interview, 1999). It was also clear that the new agency would have supranational features and that it would be subordinated to the WEU Charter (Schlieper, interview, 1999). In a so-called task force at the Institute for Security Studies, WEU, it was stated that the "Participants regretted that it had not been possible to organise as a subsidiary body of WEU. On the one hand, this has created a division in Europe's armament sector that is now difficult to overcome. On the other, OCCAR nations were obliged to launch a time-consuming ratification process and to cope with a multitude of practical problems (new security regulations, salary structures, social security of the staff, etc.)" (Schmitt 2000a, 2).

There are several options on the legal aspects of a European armaments agency if it is to be subject to the European Union. There are various independent agencies that are linked to the European Union, and their
legal status varies. Some agencies are treaty-based and others are the result
of Council regulations. There are also European bodies linked to the Commission.[21] "Its status could be fixed either in a protocol annexed to the Treaty or in a convention. These were the solutions adopted for the European
Central Bank, the European Investment Bank and Europol. If the Arma-
maments Agency followed the lines of the Central Bank, its status would
be defined in the Treaty but could be amended by a simple decision of the
Council" (Schmitt 2000a, 5). A problem with this solution is, of course,
the matter of how to handle the countries that are non-EU members but
members in WEAG (Norway and Turkey). The supranational policy on
the issue of armaments, pursued by parts of the European Commission,
has faced heavy objection from the EU member states (Heisbourg 2000).
This is in line with the national resistance to transferring sovereignty to
the European Union by abolishing Article 296 in the Amsterdam Treaty.
Another option is to create a European agency within an intergovernmental
setup within the European Union. An all-fifteen agency has, however, also
been controversial since there are deep differences between the countries
that dominate the European defense industry and those countries that have
no defense industry or only a very weak one (Heisbourg 2000).

Given the decision to incorporate the WEU into the EU, it is an open
question what will happen to WEAG and WEAO. The two organizations
have, however, been active in showing that they are willing to change to
meet the demands and requirements that are set up for cooperation within
OCCAR and LOI. By July 2000, WEAG was preparing "a new MoU
called EUROPA, which seeks to make the EUCLID system more flexible."
(Schmitt 2000c, 70).[22] Flexibility is thus the key word in WEAG's process
of change. Officials within WEAG recognize that their organizational
setups and cooperative arrangements are challenged by OCCAR and LOI
(Interviews, high officials at WEAG and WEAO September 1999). These
cooperative arrangements are sometimes referred to as restricted clubs and
elite clubs. OCCAR and LOI have moved faster than WEAG, and this is a
complication (Interviews, high officials at WEAG and WEAO September
1999). One of my interviewees argued that the principle of *juste retour* in
WEAG stands against the principles of the internal market. This principle is therefore toned down in a policy paper from WEAG in June 1999

(Mezzadri 2000). In addition, it also faces possible incorporation within the EU.

What, then, is the future for WEAG? One of my interviewees said in September 1999 that first there would be new institutions within the defense area, and then they will decide what to do with WEAG (Interview, Official, European Commission DG 1A September 1999). He outlined two major options. The first was that WEAG would be abolished when the WEU is incorporated into the EU. The second option was that WEAG would become a legal body. For now, WEAG is dependent on the WEU, but legally it is not linked to the WEU. Thus, the formal link is weak, and the WEAG secretariat has no mandate to act independently (Andrén, interview, 1999). An important reason for this is that the governments want to avoid unpleasant surprises and coups (Andrén, interview, 1999). However, from the perspective of the permanent staff within WEAG, the weak legal status is a problem and it would be much better if WEAG were to become a legal personality (Delhotte, interview, 1998). This would give WEAG a stronger position in relation to OCCAR, which in November 1998 decided to strengthen its legal personality (see above). WEAG would then become a natural base for the would-be European agency on defense procurement.

WEAO's legal situation is different from WEAG's since WEAO has its legal base in the Brussels Treaty (the WEU Charter) and can therefore exist as long as the Brussels Treaty is not changed. The commitment of WEAO, as one of my interviewees put it, will still exist even if the WEU is incorporated within the EU. So, WEAO will probably take over tasks from WEAG, whereas other issues such as procurement will be handled by the EU's community pillar (Woodcock, interview, 1999). Not surprisingly, the people within WEAG do not support this scenario. One of my interviewees argued that WEAG would not be incorporated within WEAO. This is so because WEAO was mandated by WEAG. This means that WEAG will be an umbrella organization in accordance with the latest masterplan of a European armaments agency. OCCAR and other separate initiatives will be compatible, and eventually there will be a more coherent organizational structure (Interview, High official at WEAG September 1999). Pierre Delhotte, head of the WEAG Armaments Secretariat, argued at a seminar in November 1999 that there is a "need for a forum with political dimension directed by Defence Ministers open to all European Nations involved in the creation of ESDI . . . WEAG/WEAO is the most appropriate arrangement" (Delhotte, 2000). The advantage and disadvantage of

WEAG/WEAO is thus that they comprise more countries than LOI and the EU. A problem with WEAG that is often mentioned in my interviews with people inside WEAG and WEAO is WEAG's lack of legal personality. This means above all that the permanent staff has no authority to take any initiatives but is dependent on the member states, which take unanimous decisions. These decisions are not legally binding.

Other problems within WEAG are differences between national legislation. Furthermore, the structure and scope of the defense industry underline the diversity of the participating countries and the difficulty of reaching common decisions. Some of the participating countries, that is, the OCCAR and LOI countries, dominate the production of European armaments whereas other countries have a very small and undeveloped defense industry. One of my interviewees said that the various formats of existing organizations involved in the armaments sector complicate the consensual approach on various subjects addressed by WEAG (Interviews, high officials in WEAG/WEAO September 1999). The decisions within WEAG are often the result of a minimal solution, that is, they represent the lowest common denominator (Interviews with high officials in WEAG/WEAO 1999). Other problems concern the disparity of the participating countries in WEAG. WEAG's diversity is also its strength. WEAG members that do not participate in LOI and OCCAR represent more than 40 percent of the defense market (Interviews, high officials in WEAG/WEAO September 1999). WEAG has also been instrumental for the defense ministers since the door has been closed in the European Union. This argument, of course, is weak in a situation in which the defense ministers now can formally meet in the EU. One of my interviewees pointed out one of the paragraphs from the Cologne Declaration, which supports the notion that armaments issues "belong" to WEAG. The paragraph refers to the defense industrial and technological base and harmonization of military requirements that are crucial components in the work of WEAG (chapter 3). However, WEAG is not mentioned in the text, and this is open to interpretation (Interviews, high officials at WEAG/WEAO September 1999).

To sum up, there are several organizations or a combination of organizations that can form the basis of a European armaments agency. The perceived need to create such an agency has increased during recent years as the formation of a European defense policy has taken shape. Indeed, the question of procurement stands out as a crucial component of a European armaments policy, which, in turn, is an important part of a European defense policy (EDP). As Francois Heisbourg has put it in a report from the WEU Institute in Paris, "To the extent that EDP implies an autonomous military capability, the Europeans cannot afford to be solely, or even principally, dependent on the U.S. defence industry, whose primary loyalty (in terms of production and overall service) is naturally to its main customer, the US government" (Heisbourg 2000, 101).

The Challengers and the Traditional Organizations

The creation of OCCAR, the LOI initiative/framework agreement, and the decision to incorporate the WEU into the EU have initiated an active and complex process of interpretations by the organizations of the organizational consequences. This organizing process is not about technical issues. On the contrary, the rivalry and competition between the organizations concern fundamental political questions. How politically acceptable is flexible cooperation in defense matters? Can a few politically and industrially strong countries decide the rules of the game for European cooperation on armaments (for example OCCAR and LOI)? Will the cooperation lie outside the traditional organizations, and can the EU handle an issue that concerns countries that are not EU members? Which role will the EU's rules on the internal market, competition, etc. within the community pillar have in regulating European cooperation on armaments?

It is clear that the traditional organizations—the WEU, WEAG, and WEAO—are being challenged by the "pretendents" (OCCAR and LOI). The traditional organizations must deal with the challengers and are under great pressure for change. The pressure from OCCAR and LOI is made possible due to the general political process on the issue of armaments that has opened up new possibilities for organizing cooperation. It's obvious that the decision to abolish the WEU has given the opportunity for criticism. The WEU structure represents the Cold War, and the question is how much of the old structure will remain in the new structure of European cooperation on armaments.

There are new emerging forms of domination and authority structures that are both formal and informal. OCCAR and LOI are based on cooperation principles and were initially very loose organizations based on voluntary agreements and network relations. Gradually these organizations have incorporated a more legal setup and agreed on legally binding agreements. The emphasis on the importance for an organization to have legal personality suggests that this is crucial to gain respect and to become an authoritative and legitimate actor. Paradoxically, the informal and flexible way of organizing is at the same time regarded as something positive and valuable. This seems to be the case with the LOI working groups, which often directly criticized the inflexibility of WEAG and WEAO. OCCAR and LOI float freely in the sense that they do not belong to either the EU or the WEU. They are part of the two organizational fields, market and defense, although defense considerations seem to predominate. Nevertheless, it can be argued that OCCAR and LOI are important organizations in the new organizational field that emerges.

Scope and Depth of Organizational Interaction

The previous chapters and the sections in this chapter show that there are several organizations that deal with the issue of armaments. They compete over which organization, and thus regulatory framework, the European cooperation on armaments should adopt. The organizations are also dependent on each other. Indeed, it is difficult to imagine European cooperation on armaments without a regulatory and an organizational setup that do not consider both the market frame (the European Commission and the internal market rules, etc.) and the defense frame (the EU's second pillar). Another matter is where the emphasis will be. Even POLARM recognizes the importance of the market implications of the issue of armaments but is reluctant to hand over more power to the Commission.

In this section I continue to analyze the basis for the cooperation between the organizations within the two fields. Do they communicate with each other and what kinds of interactions are taking place between the organizations? In addition, what are their attitudes toward each other?[23]

In my interviews I have been struck by the fact that people inside the European Commission, POLARM, the LOI initiative, WEAG/WEAO, and CNAD in general seem to have very limited contact with each other, or at least that they do not willingly admit that they do have contact. They have also occasionally been rather ignorant of the activities that are taking place elsewhere.[24] However, it could be that some of the people that I have interviewed are either not aware of this interplay between the organizations or they are being careful in discussing a politically sensitive matter with a stranger.[25] Nevertheless, the lack of interest and knowledge that I have sometimes encountered is rather unexpected since there is a rather limited number of people who take part in the organizations dealing with European cooperation on armaments. The memberships in these organizations sometimes overlap, that is, the same people sit in two or three organizations and discuss the issue of armaments. This is the case at the political level as well as among the officials. The discussion within WEAG encompasses both EU members that are NATO members and EU members that are nonaligned and governments that are NATO members but not EU members. The work within CNAD concerns NATO members primarily, but PFP governments are also included. Some of these participants also take part in WEAG and WEAO. Four of the European governments have started OCCAR, and they are simultaneously participants in CNAD and WEAG/WEAO, LOI and the EU. These four countries plus two more countries have initiated LOI. These six countries also take part in the discussion within the EU with countries that are not participants in the OCCAR and LOI. The latter countries take part in the work by CNAD and WEAG/WEAO.

On the industrial side, the overlapping membership sometimes results in amusing situations. NIAG (the industry group within NATO) is reluctant to come to WEAG and talk about European cooperation on armaments even though several NIAG participants are also participating in EDIG, which is taking part in the discussion within WEAG (Woodcock, interview, 1999).

This is clearly a paradox: that the representative of one organization does not acknowledge that he or she has contacts with representatives of other organizations when they are simultaneously taking part in the same circles. How can this paradox be explained? First of all, who are "they"? The interviews were conducted with officials of the Commission and WEAG/WEAO and with a few national representatives of POLARM, LOI, and CNAD. The discussion of the contacts between the organizations is in some cases also based on official documents. In the previous section on OCCAR, LOI and the would-be agency, it is shown in the documents that there is an awareness of the competition between the organizations, and that the people inside the organizations do monitor each other and present policy responses (for instance, consider WEAG's attempt to appease the LOI participants). However, the way organizations present themselves is important and a sensitive issue in a situation in which there is a great uncertainty over the European organizational setup on armaments. One of my interviewees within the Commission stated, "we don't know who our big father/mother will be."

NATO-WEAG and NATO-EU

The informal link is strong between the WEU and WEAG since the majority of the participants in WEAG are also members of the WEU and of NATO. This is also the case between WEAG and NATO since the majority of the participating countries within WEAG are also NATO members.[26] The cooperation between WEAG/WEAO and NATO, however, is rather limited, even though some of the CNADs (NATO) are also NADs within WEAG. The reports from NATO on the formation of a NATO Armaments Community show that NATO is more occupied with internal coordination than with external organizations (chapter 3). The external world, that is, other European organizations dealing with armaments, is hardly ever mentioned in the CNAD reports. In one of the reports, however, it is stated that the development of ESDI requires cooperation between NATO and European bodies—"closer cooperation in the field of armaments with the Western European Armaments Group and Organization (WEAG/WEAO) should be actively pursued" (NAC 1997, 6). However, the panels of WEAG, especially Panel I (see chapter 3), have contacts with CNAD. It is stated in the presentation of the work of Panel I that it "has to avoid any duplication of effort between cooperation programmes in the NATO framework and its own programmes."[27] Furthermore, in 1994 CNAD agreed on new practical cooperation measures with WEAG, "providing a means of expanding the dialogue on transatlantic armaments issues between European and North American allies" (Naumann 1996).

The formal and informal links between NATO and the European Union are very limited. These limited interactions concern both the organizations that are dealing with the defense frame (NATO and the second pillar of the EU) as well as the civilian organizations (from the community pillar) and NATO. The general question of the relationship between NATO and the EU has traditionally been politically controversial. In the declaration by the European Council in Cologne 3-4 June 1999 it is stated that there is a need to ensure consultation and cooperation between the EU and NATO. The building of a defense policy/capacity within the European Union suggests that this dialogue will be intensified. This has indeed been further proposed at the EU summit at Feira, Portugal, in June 2000 (European Council 2000, Presidency Conclusions, Annex I). In the day-to-day work on the issue of armaments the contacts and trust between NATO and the EU seem to be more difficult to realize. The European Commission has for many years stated that it is important to have regular exchanges of information between NATO and the European standardization bodies concerning the defense industry (COM, (97) 583, final). The European Parliament has also emphasized the need for better coordination among the various activities related to armaments (A4-0482/98). Even though there are some contacts between the community pillar and CNAD, they generally are limited.

An overall impression is that NATO's work on the issue of armaments concerns a whole range of questions that are linked to the specifics of NATO and that this track is moving rather slowly, whereas the more European track, with WEAG, LOI, EU, etc., is moving faster. There is also the view that the activities within NATO and the EU are in competition with WEAG and that many member states in WEAG seem to pay more attention to NATO and the EU than to WEAG (Interviews, high officials at WEAG/WEAO September 1999). However, NATO's work with standardization is also regarded as more successful than the work within WEAG (Interviews, high officials at WEAG/WEAO September 1999). The PFP cooperation reactivated NATO's work with arms standardization. NATO thus seems to be the most important organization when it comes to military capacity. Furthermore, it is also quite clear that some of my interviewees see the NATO work as an American track and that this American work on the issue of armaments stands against a European cooperation process. The United States, some believe, tries to control WEAG (Interviews, high officials at WEAG/WEAO September 1999).

WEAG/WEAO and the European Union

A closer relationship between the EU and the WEU is emphasized in the Amsterdam Treaty and the declaration by the Western European Union on the role of the WEU and its relations with the European Union, adopted on July 1997. These documents are explicitly mentioned in the Commission's draft for a common position on a "European Armaments policy" in the communication from November 1997. In Erfurt in 1997 the defense ministers within WEAG decided to establish a special group —the ad hoc Armaments Working Group—to analyze the options for a European armaments policy. The group consisted of people from the EU and WEAG. They were asked "to identify the points in the report which warrant further examination within the European Union Framework, make recommendations for further action within the Community framework or within the framework of Title V TEU, and if appropriate, list suggestions for specific measures, without prejudice to the Commission's competence under the EC Treaty" (POLARM 1998b).

In the day-to-day cooperation, the relationship seems to be rather tense and undeveloped between WEAG, the European pillar of armaments, and POLARM (the ad hoc working group in the Council concerning armaments). A sensitive issue is how the work of POLARM should be linked to WEAG. The two bodies are well aware of how the issue of armaments requires cross-border cooperation between the WEU and the EU. The POLARM group often states that the issue of armaments concerns the EU as well as WEAG. The work between WEAG and POLARM runs in parallel. "It is not coordinated and there is a risk of duplication. Informal exchanges of information between POLARM and WEAG take place on an ad hoc basis through the EU and WEAG Presidencies" (POLARM 1998b, 2-3; POLARM 1998e: 3; POLARM 4 June).

In several POLARM documents it is stated that to enhance transparency and the effectiveness of coordination of work in the EU and WEAG, "exchanges of information should be placed on a formal basis" (POLARM 1998b, 3; POLARM 1998e, 3). These documents on enhanced cooperation between the EU and WEAG have been produced in consultation with the WEAG presidency (POLARM 1998e, 3). The proposed contacts consist of regular exchanges of information on on-going activities in each organization and joint sessions on relevant topics, and they should take place between the Secretariat-General of the Council, the presidency of POLARM, the Armaments Secretariat of WEAG, the panels of WEAG and the National Armaments Directors (POLARM 1998e, 3). It is unclear, however, how these contacts should be organized and institutionalized. In the case of WEAG, it might be asked whether there is anyone at all who can represent this transnational organization. The permanent staff is very limited and its authority to speak for the member states is not self-evident. Informal exchanges of information take place on an ad hoc basis through the EU and WEAG presidencies, but there is a clear unwillingness to formalize the contacts between POLARM and WEAG. The proposed arrangements for formal contacts between the EU and WEAG, for instance, are not supported by the Greek delegation, which wants informal and ad hoc contacts on a case–by–case basis (POLARM 1998b). The Commission representative in the talks between POLARM and WEAG has, however, underlined the inter–pillar aspects and that this motivates closer cooperation between the organizations (POLARM 1998b). In a POLARM document from June 1998 the reference to formal exchanges of information is replaced by regular exchanges of information (POLARM 4 June 1998: 2).

The relationship between WEAG and the Commission seems less tense. Parts of the Commission, especially DG IA and III, frequently meet Panel III of the WEAG. On a higher level this means contacts between "the WEAG Panels through the Chairman of National Armaments Directors by the EU Presidency and the Commission" (POLARM 1999b). "The Commission plays an active role in EU discussions of armaments and makes proposals according to its responsibilities. On matters of mutual or complementary interests, the Commission and WEAG maintain contacts between the representatives of appropriate Commission DG's and the relevant WEAG panel, which they are seeking to intensify. On matters of mutual or complementary interest the Commission and WEAG maintain informal contacts and in this context have periodic exchanges of information" (POLARM, Annex, 4 June 1998). When the Commission presented its proposal in November 1997, WEAG contacted DG III since six or seven issues of the action plan already were considered within WEAG (Linton, interview, 1998). They worked together during 1998, but, due to the general turbulence of the Commission in early 1999, the work was stalled (Linton, interview, 1998).

The informal contacts are thus extensive between the Commission and WEAG. EDIG also takes part in the formal meetings with Panel II and the Research Cell. Indeed, in March 1998 the so-called SCITEC report was presented in which EDIG participated; it was titled "Building on Success, WEAG Science and Technology Strategy."[28] The aim of the study was to develop a WEAG science and technology strategy. The study was commissioned by WEAG Panel II and was undertaken during 1996 and 1997. The NAD's have approved the report. The study identified technologies that are expected to have a profound impact on defense capabilities, etc. "A common WEAG doctrine would have a significant impact on cooperative programmes" (SCITEC 1998, 2). Interdependency is the key word: "In order to reduce duplications and associated waste of resources, an increased cooperation is needed, resulting eventually in a situation of interdependence, where each European state will contribute to the European technology base in some areas and rely on other States for other areas" (SCITEC 1998, 5). It is thus necessary to establish a European strategy and closer a relationship between defense and civil research programs at the European level.

An Armaments Network

The contacts between the traditional organizations from the two fields seem to be rather limited, at least when it comes to formal links. This is not surprising given the overall logic that permeates this book. However, the European Union (both the first and second pillars) has a rather tense and difficult relationship with NATO and WEAG on the issue of armaments. This is more surprising since the second pillar of EU belongs to the defense field as NATO and WEAG do. One general explanation for these tensions is given by De Spiegeleire, who argues that the "WEU and NATO share a political-military culture that remains alien to the European Union" (De Spiegeleire 1999, 88).

The relationship between the traditional organizations and the newcomers also seems to be rather limited. There are, however, quite comprehensive informal contacts between LOI and WEAG (Törnquist, interview, 2000). Furthermore, the relationship between the activities within the EU, especially the General Directorate of industrial affairs in the European Commission, and the work within LOI, has not been very intense. In fact, when the executive committee presented the second LOI report to the six defense ministers in July 1999 (released in September 1999), I was the first person to give the report to the officials in the special unit at the DG on industrial affairs dealing with European industry! The analysis of LOI and its working groups showed that LOI hardly ever mentions the European Union, WEAG/WEAO, or NATO (with the exception of the Research and Technology group). The traditional organizations, especially WEAG/WEAO, are on the other hand more interested in having contacts with LOI and OCCAR. In the masterplan of WEAG/WEAO, for instance, it is stated that the would-be agency, EAA, should be compatible with OCCAR.

In April 1999 the Commission initiated a meeting with LOI, OCCAR, WEAG, EDIG, etc.[29] The meeting was intended to promote a climate of confidence among the various actors involved and to create a better knowledge of the work they currently conducted (Gunnarsson, interview, September 1999). Although the turnout proved to be rather low, it was a first initiative to bring all the actors together in an effort to improve the communication and coordination among them. The meeting was clearly an attempt by the Commission to try to take back the initiative on the issue of armaments from the six governments within the framework of LOI. The meeting also reflected the increased concern that the lack of communication and coordination between the various organizations will fail to enhance Europe's competitiveness toward the United States (Interviews, high officials at the Commission September 1999).

There are contacts between the organizations that are not taking place through the ordinary channels. Several policy research centers/institutes have been very active in bringing the issue of armaments to the political agenda. These centers/institutes have, on a regular basis, met representatives from the European defense industry, officials from the Commission, officials from NATO, WEU, WEAG, and WEAO. It is through these centers/institutes that participants from various organizations have met in a context that is less political than would have been the case with interorganizational contacts.

The restructuring of the European defense industry has been discussed in a seminar series, a so-called working party, at CEPS—an influential research center in Brussels. The list of the participants is impressive, ranging from representatives of leading defense industries in Europe to high officials within WEAG/WEAO and the European Commission (CEPS 1996; 1999).[30] The European Commission has provided financial support for the work by the working party and "participated in our work from three different angles: competition, industrial policy and the Common Foreign and Security Policy. Its important communications on the European Aerospace Industry and on Defence Related Industries of 1997 figured prominently in our proceedings" (CEPS 1999, preface). The members of the working party met regularly over a two-year period and were especially focused on the creation of European multinational forces. The group has been led by Dr. Willem van Eekelen, former Secretary General of the WEU and CEPS Senior Research Associate (CEPS 1996; 1999). Evidently the participants within this working party are a rather homogeneous group. From the perspective of an outside observer it is quite clear that they know each other well and that they formulate proposals for future action. In my research and in my effort to identify people who work with the issue of armaments I soon realized that these people were to be found in this working party. Everytime I found an interviewee, I discovered that the person was on the list of members of the working party or that the person was knowledgeable about the work of the group. Several of my interviewees also stressed the importance of taking part in the meetings with the working party.

The rationale behind the formation of a special working party and its holding several meetings/seminars is not only to inform each other about various events but also to bring all the relevant actors together in an effort to foster a common policy on the future of the defense industry. This conclusion can be drawn from the fact that the reports from the working party consist of many policy recommendations. Some of these recommendations have already been mentioned in chapters 3 and 4. Furthermore, it is also obvious that the chairman of these meetings has pursued a common European defense policy.[31] The working party and its participants have therefore discussed the requirements for multinational forces in which one important precondition is the standardization of armaments. In a concluding report from the working party in June 1999 several policy recommendations are presented that concern both policy questions and also interorganizational activity. Contacts

> and relations between OCCAR and the European Commission should be strengthened. Even more generally, cooperation between the EU, POLARM, WEAG and EDIG should be institutionalised. The Commission Action Plan contains at least five priority areas that are similar to WEAG priorities. Coordination in these areas could be established immediately. Clear structural developments are needed to make WEAO and OCCAR compatible. (CEPS 1999, 2)

Another institute is the Institute for Security Studies, Western European Union, in Paris.[32] A special task force (on the European Armaments Sector) consisting of researchers, high representatives from the European defense industry, and various decision makers has been discussing the future of the European defense industry. It is obvious that the meetings have functioned as more than just a way for the participants to get information on the complex political process on European cooperation on armaments. The group also makes specific recommendations that it perceives as urgent and important. It is, for instance, in favor of a legal strengthening of OCCAR (Schmitt 2000a). The task force has also discussed a division of labor between the existing organizations—WEAO for research and technology, OCCAR for program management, and EAA for in-service support. "From an economic point of view, an exclusive arrangement between the six major arms-producing countries might be possible. From a political point of view, however, a comprehensive organisation including all European countries would be preferable" (Schmitt 2000a, 4). On the issue of a would-be agency, the task force excluded the options of a body depending on the WEU and a holding of existing organizations whereas the options of a body of the European Community or the European Union were more realistic (Schmitt 2000a).

It is of course difficult to establish in what ways these policy centers have had any effect on the European policy on armaments. It seems clear, however, that these centers, especially CEPS, were important meeting places in the early phase of the policy formation. Only two of my interviewees have argued that the centers had no effect at all. One of these persons took part in one of the centers and discussions whereas the other person did not. They both argued that these centers did nothing but talk and talk. I would argue that talk is very important in the policy process, especially in its early phases. Talk and discourses enable new ways of defining and framing issues and create frames that are taken for granted.

In the case of CEPS and the WEU Institute for Security Studies, nothing new came out of the discussions as far as the documents are concerned. In the CEPS reports, the frames, both the diagnostic and the prognostic, are clear and pronounced, resembling the communications from the Commission. My interviews do not confirm that any dramatic change concerning how to perceive the issue of armaments took place within the discussions in the centers either. What the interviews do confirm, however, is the notion that these centers, especially CEPS, took an early initiative to gather various organizations in a situation that was characterized by organizational fragmentation and disjointedness. They gathered organizations from different fields and started a dialogue that continued for several years. CEPS and the WEU Institute for Security Studies provided informal and less politically constrained fora than the EU or WEAG could offer. They were also able to gather people from organizations that had tense relationships. The CEPS working party functioned as a meeting place for private and public actors who worked with the issue of the defense industries and other related issues. Within this network they not only exchanged information but also fostered a common perspective on how to restructure the defense industry and in what ways the political decision makers could be approached.

My empirical analysis suggests that representatives from the Commission, WEAG, the industrial organization EDIG, the European industry, and researchers form a loose defense-industrial network. It was not an epistemic community; that is, it did not contain a group of experts and technocrats with collectively shared values and consensual knowledge. Although it is possible to identify certain components of collectively shared values and consensual knowledge, the network seems to be more heterogeneous than an epistemic community approach requires (Haas

1992). In fact, there seem to be several networks that are held together by the common view that there was a need to develop a European policy concerning the defense industry and to create European cooperation on armaments. The relationship between the defense and market frames had to be strengthened to create such a European effort.

The empirical material also shows that the people inside the organizations are not always informed of each other's organizational activities. This lack of knowledge or interest can be interpreted as an important component in the process of organizing between the participants in the fields. They monitor each other but they do not develop any patterns of interaction because this would mean that they would grant each other an active role in the field in the making.

Notes

1. www.premier-ministre.gouv.fr/fr/p.cfm?ref=4235&d=361. 6 February 2000. Translation by the author.

2. The new number is twelve in the consolidated Amsterdam Treaty.

3. A special working group is also discussing export policy—COARM.

4. According to the German delegation, POLARM should deal with the following topics: intracommunity transfers, public procurement, competition policy, and principles for market access (POLARM 1998g).

5. Seminar, 6 December 2000 with Swedish POLARM and COARM representative Stefan Nilsson at the Swedish agency for strategic export.

6. The role of those DGs within the Commission who deal with CFSP issues seems to be strengthened vis-à-vis the more industrial and market-oriented DGs. Although the tension between the DGs has been reduced due to the reframing in the communication in November 1997, it is also clear that the DGs still compete with each other.

7. Article 297 (formerly 224): "Member States shall consult each other with a view to taking together the steps needed to prevent the functioning of the common market being affected by measures which a Member State may be called upon to take in the event of serious internal disturbances affecting the maintenance of law and order, in the event of war, serious international tension constituting a threat of war, or in order to carry out obligations as it has accepted for the purpose of maintaining peace and international security."

8. Article 133 after the Treaty of Amsterdam.

9. See also COM (92) 317 Final and SEC (92) 85 Final.

10. The Netherlands has applied for membership. Belgium has been a candidate since 1998, and Sweden officially showed its interest in joining the organization

in January 1999.

11. Programs managed by OCCAR include the Hot, Milan, and Roland missiles programs, the Tiger helicopter program, and the Cobra counterbattery radar program. For an overview of the programs within OCCAR see *L'Armement* 1998 and Mezzadri 2000.

12. The problem of *juste retour* deals with the question of collective action for private goods. States seek assurances that they will get a fair return on their investment in the cooperation (Sandholtz 1992). National juste retour means that "each member expects a balance between contribution and rewards, between what it puts into and what it receives from the organization" (Sandholtz 1992, 28). A global juste retour means that this balance between contribution and rewards is less direct and met in the long run.

13. CEPS 1999; Interviews with Lars Fagerberg (1999), Dag Törnblom (1997, 1998), Stefan Törnqvist (1998, 1999, 2000), Tommy Ivarsson (1998), Åke Svensson (1998) and Laurie Frier (2000).

14. Seminaire vers une culture europeénne de défense et de securité, www.defense.gouv.fr/actualites/communiques/d100700/100700.htm. 6 February 2000, see also Britz and Eriksson 2000)

15. Representatives from the industry have participated on various occasions but have not been formally part of the working groups.

16. TDC stands for Transnational Defense Company.

17. OCCAR is not discussed since it is not involved in harmonization of military requirements.

18. Annex to letter from the Chairman WG HMR dated 15 June 1999.

19. The EU Code of Conduct on Arms Exports was formally adopted by the EU Council of Ministers as a legally nonbinding Council Declaration in June 1998. "The Code's significance is partly in the elaboration of guiding principles to be taken into account when considering arms export licence applications and partly in the operative provisions it establishes for information exchange and consultation. The Code sets out minimum levels of restraint and allows member states to operate more restrictive national policies if they so wish" (Davies 2000, 17).

20. A representative from the Swedish Foreign Ministry took part in the work of the executive committee and his task was to monitor the linkage between LOI and the work within POLARM and COARM.

21. There are several treaty-based independent European agencies, for instance the European Central Bank (Treaty of the European Union). The European Environment Agency is an example of an agency that was established as a result of a Council Regulation (1210/90/EEC, OJ L 120) May 1990. The Joint Research Centre is an example of a body linked to the Commission.

22. EUCLID is an RTD cooperation program.

23. The analysis is the result of a number of interviews with people inside these organizations and the result of my reading of various documents from

the organizations in question. The empirical material is rather limited and does not cover *every* kind of interorganizational interaction that is taking place. It is, however, my belief that I have covered important activities between the organizations and their attitudes toward each other.

24. On a few interview occasions I have presented documents from other organizations that the interviewees wanted to copy since the documents were new to them! These people were high officials and in the center of the policy process on armaments, for instance a senior official within the General Directorate for external relations, with responsibility for the issue of armaments.

25. In an earlier study on the European Commission one of my interviewees told me that her superior had told her to not talk to anybody! (Mörth 1998).

26. Countries that are either members, associate members or observers in the WEU are: Belgium, Denmark, France, Germany, Greece, Italy, Luxembourg, Netherlands, Norway, Portugal, Spain, Turkey, the United Kingdom, Austria, Finland, Sweden. Most of these are also members of NATO.

27. www.weu.int/weag/eng/info/weag.htm, 16 November 1997.

28. SITEC stands for Study Team established by Defence Ministers of Nations of the Western Armament Group with the participation of the European Defence Industry Group.

29. The Commission has also arranged informal meetings between the defense and industry ministers during 1998 (Interview with Senior Official, European Commission, September 1999).

30. In the concluding report from 1999 the chairman of the group, Willem van Eekelen, mentions prominent participants in the work. "The Western European Armaments Group was represented by its chairman and the head of its secretariat and the WEU by the deputy head of the Research Cell. The WEU Planning Cell and the WEU Institute for Security Studies in Paris were also represented. From NATO, the group was joined by members of the Defence Support Direction, including the Director of Armaments Planning, and by the Head of the Capabilities Co-ordination Cell of the International Military Staff. The Armaments Counsellors of the delegations of Germany, France, the Netherlands and the U.K. were regular participants. We greatly appreciated the active participation from the side of the European defence industry, notably British Aerospace, Celsius, DASA, Dassult, FIAT and SAAB. We also acknowledge the financial contribution of the Netherlands Ministry of Defence" (CEPS, 1999, preface).

31. The former General Secretary of WEU, Wilhelm van Eekelen, has functoned as chairman of the working party and was until 1999 a senior research associate at the Centre for European Policy Studies, Brussels.

32. The Institue for Security Studies is now affiliated with the EU.

Chapter 6
The Organizational Field: Frames, Authority, and Interaction

What are the theoretical and empirical implications of the three consolidating processes of the formation of a European organizational field? The processes are:

1. Reframing of the issue of armaments
2. New forms of formal/informal domination and authority structures
3. Scope and depth of organizational interaction.

Chapter 5 shows that both parallel and coordinated activities are undertaken between the organizations from the two organizational fields —market and defense. The organizations from the two fields have partly different histories and origins, but their roads seem increasingly to cross. The borders between the organizations are blurred. This can be explained by the fact that the end of the Cold War has opened possibilities for multiple interpretations of issues that have traditionally been exclusively defined by interpreters within the sphere of anarchy (NATO, the WEU, defense ministers, etc.), or by interpreters within the sphere of interdependence (industrialists, officials within the European Commission, ministers of industrial affairs, etc.). Various issues such as the issue of armaments, or concepts such as security, are no longer easy to categorize, and, above all, it is increasingly difficult for actors to legitimate an order in which the issue of armaments belongs exclusively to either a market-making organization or to a defense-oriented organization.

One important aspect of the traditional separation between the two fields—market and defense—is that the people from the organizations have very seldom interacted. This means that conflicts related to cross-issues have been avoided. Thus, by keeping the two fields separate, struggles over how to frame an issue and over who owns the question have been kept within the two fields and their respective organizations. The recent political development has allowed a more open discussion of how a European armaments market is dependent on the existence of a European security and defense policy, and it has become clear that the European security and defense identity will have no substance without an industrial and technological capacity. Putting the frames together and arguing that the issue of armaments needs both market and defense considerations opens a discussion about, and conflicts over, how to link the two frames; these are conflicts that were earlier kept within the two fields, or that were hardly discussed. How can the protection of national defense interests (Article 296) be combined with an internal market for armaments? How will the governments handle the issue of security of supply in a situation when there are private European companies? How can organizations that belong to different paths of the European integration processes cooperate and communicate with each other on these questions?

The European Commission has forcefully argued for an interlinkage between the market and defense frames. The LOI governments, the Council of Ministers, WEAG/WEAO, and the defense industry have also argued that the defense industry development is linked to both frames. Another question concerns the regulatory and organizational implications of such an interlinkage. It is clear that the six governments within the LOI are not prepared to communitarize the issue of armaments but intend to keep the issue within the intergovernmental path of the integration process. It is unclear whether the reframing of the issue of armaments in the Commission can be interpreted as a pragmatic "escape route" (Héritier 1997) in order to move on in the political process, or if it is an expression of a more profound change in the rules of the game. This study does not give any definite answers to that question other than that there are clear signs that the issue of armaments is reframed and that this can be seen as a way out from a political and organizational deadlock. The elaborate communications from the Commission, however, suggest that it challenges the traditional separation of the two projects of European integration regarding the issue of armaments. This is a field in the making and its autonomy is therefore weak.

The power struggle over the question of which organizational field the issue of armaments belongs to is most evident within the Commission, but the tension between the two frames is also evident in the working group in the Council (POLARM) and in the European Parliament. One important conclusion in the study is that the conflicts over the issue of armaments do not take place only between organizations from different fields but also within organizations (POLARM, European Parliament) and within organizations that belong to the two fields at the same time (Commission). The tensions within and between organizations concern fundamental conflict dimensions in European politics. The most obvious dividing line is the one between a supranational decision-making structure and an intergovernmental decision-making setup. Another important question concerns the relationship between European cooperation and the United States. In the market field, the Commission has revitalized an old discussion of how Europe is lagging behind the United States. The theme of the importance of a strong Europe is also a major one in the defense field. However, the European experience in the two fields is interpreted differently. In the defense field the EU is striving to acquire the strategic capability worthy of a great power. There is no aspiration to become the equal of the United States—for instance, in nuclear weapons or aircraft carriers—but to have the independent capability for robust military crisis management even at great distances from Europe proper.

In the market field, the Commission, especially the market-oriented part of the organization, has successfully marketed the image of Europe's defense industrial deficit toward the United States that has triggered activity within the government-led organizational activities, the Council of Ministers, and LOI. In this field the similarities are perhaps even greater with respect to what the EU and Europe want to become in comparison to what the United States already is. The main mechanism behind a consolidation within the market field, and behind the emerging European organizational field on armaments, is mimetic and is thus based on imitation of the United States in a time that is perceived as uncertain and turbulent. The coercive mechanism behind a common frame conceptualization is less salient in the empirical analysis. However, the potential for legislation in the European Union on the issue of armaments (the rules on the internal market and the second pillar) is high. The cases in the European Court of Justice also suggest that coercive rules will increase in importance with respect to how the issue of armaments is framed.

The uncertainty regarding the future of WEAG/WEAO is an expression of how the traditional frames and organizational setups are put into question. This situation of uncertainty creates turbulence and possibilities for new ways to frame and organize European cooperation. OCCAR and LOI not only are given the opportunity to change the traditional organizational setup, but they have also created a political turbulence and preconditions for change. Furthermore, technological and RTD issues are becoming more important in NATO and in WEAG/WEAO, and defense issues are incorporated into the European Union. The Cologne process suggests that the EU will be more focused on military tasks. Furthermore, the rhetoric within the market-based European Union not only conveys the message that the European countries are interdependent; but also emphasizes the notion of competition vis-à-vis the United States—the anarchy rhetoric. In NATO, which is based on the logic of anarchy, the message in various reports is the notion of the increased mutual interdependence among countries. Thus, the traditional rhetoric within the sphere of interdependence is moving into the anarchic organization, whereas traditional anarchic rhetoric is becoming a common feature in the interdependence organization.

There are overlaps between the traditional and the new forms of organizational activities. There are also important differences. One difference is that the organizational activities that lie outside the EU, WEAG, and NATO concern fewer countries. Thus, a few countries have loosened themselves from the political and difficult process within the EU and the WEU and have started to discuss the issue of armaments on their own. LOI set loose much of the political energy that had been locked in the EU and WEAG. It has brought about a smoother decision-making process involving fewer countries (those who dominate the European defense industry) and thus a more homogeneous group of countries. These countries are also less politically constrained in their cooperation than would be the case if the discussions had taken place in the Council, in which unanimous consent is the decision rule (CFSP). The framework agreement within the LOI initiative is interesting since it lies outside the EU and is not backed by the hierarchical authority of a formal organization. The national commitments are, however, extensive, which suggests that the agreement will have a profound impact on European cooperation on armaments and on national defense industrial policies (Mörth 2003).

The organizations in the emerging organizational field try to differentiate themselves to reduce competition and to establish a monopoly over a particular subsector of the field. OCCAR, WEAG, and the EU all struggle to get the would-be agency on armaments under their roofs. The constant marketing between the participants in a field reproduces the structure of the field and makes it a work in motion. The new forms of cooperation that are being presented by OCCAR and LOI suggest that there will be new forms of domination and authority structures that are based on informal network relationships and voluntary rules rather than on the traditional way of governing (formal hierarchical relations and coercive rules). This means that the traditional division and conflict between organizations that are based on two different logics in the international system —anarchy and interdependence—are less salient in the study. A more pronounced conflict, instead, is that between the established organizations and the challengers, which are both market and defense related. OCCAR and LOI (the pretendents) are interested in discontinuity and rupture, whereas WEAG and WEAO (the dominants) are more interested in continuing the Cold War organizational setup. They were born out of the Cold War logic even if they were created very late in that period.

Furthermore, there also seems to be a general understanding among the organizations, especially by the European Commission and EDIG, and in WEAG's report on a masterplan for a European armaments agency that the various organizations need each other. The question is, however, whether the organizations are integrated or if we are dealing with a constellation of organizations that are loosely connected. There is no simple answer. The interaction between the participants in the field partly constitutes a network. A limited number of people meet each other in various organizations. They seem to share a general understanding of the need for European defense industrial responses toward the United States, both in technological/economic terms and in terms of military crisis management. An important mechanism behind a common frame on the issue of armaments is thus a type of normative isomorphism. A similar educational background is not the primary tie that binds people in the network together. Some of them, for instance, the military staff and the officials within the Commission, have very different educational backgrounds. Instead, they seem to have developed a common understanding of the issue of armaments by way of years of frequent meetings, contacts, and other interactions. Thus, increased interaction between various parts of organizations could lead to a common frame on the issue of armaments. It is also obvious that the European Commission's effort to reframe the issue of armaments has started intense activity and interaction between the participants in the field. The political breakthrough for discussing armaments as a complex issue that includes both frames creates possibilities for new types of alliances and patterns of interaction.

The relationship between the three consolidating processes is complex. They seem to reinforce and enhance each other. The source of change is therefore not easily determined, although it is quite obvious that the interaction between participants from various organizations led to a more common understanding of how to deal with the issue of armaments. It is also quite clear that the Commission's work to reframe the issue of armaments took place in rather an early stage of that organizational interaction.

Chapter 7
Organizing European Cooperation on Armaments: Conclusions and Discussion

Issues do not speak for themselves. They must be interpreted and given meaning. In this study these interpretations have been analyzed as frames that are based on different organizational fields. These fields are held together by regulative and constitutive rules. The empirical study in chapters 3 and 4 showed that issues travel between different organizational fields and political contexts. The issue of armaments was given different contents and meanings depending on the field within which it was discussed.

Attempts to harmonize armaments procurement in Europe have a long history. Until the mid-1960s the political initiatives came primarily through NATO. It is striking, however, that NATO has achieved limited coordination in practice in this area. A European voice on armaments and defense industry matters was established in the mid-1970s with the standing up of the Independent European Program Group (IEPG), the predecessor to WEAG in 1992. WEAG marked the WEU's first serious involvement in the armaments sector, but its future is uncertain in a situation where the WEU will be incorporated into the EU. In the late 1990s and in the early part of this century, the defense frame on the issue of armaments was activated in the EU due to the incorporation of defense issues. In various intergovernmental EU documents it is stated that the end of the Cold War requires military crisis management and that these new types of military operations make interoperability and standardization of armaments necessary. European cooperation on armaments should, according to the Council of Ministers and the European Council, be based on the second pillar of the EU and on intergovernmental decision making.

The issue of armaments had, however, already been part of another organizational field—the market field—in which the issue of armaments was treated from a technological and economic competitiveness perspective. This field is based on the notion that Europe is technologically lagging behind the United States. The Commission has, together with the industry, and through special journals and policy centers such as CEPS, created a political crisis awareness pointing to various economic, industrial, and technological threats from the United States. An image of a Europe in disarray is presented that legitimizes a stricter interpretation of Article 296 in TEU and the protection of national security interests. Important economic and technological values are at stake that make it necessary to think in European and in global terms. The nation-state as the important territory for technological innovations was replaced by the notion of a European territory. In the 1980s the civilian high-tech industry was consolidated into large transnational companies. The EU developed more coordinated RTD and industry policies. In the 1990s the turn has come to the defense industry. Thus, the ongoing liberalization of the national defense industries is part of the liberalization trend in the European Union that started with the creation of the internal market. This market integration entails the gradual harmonization of standards and the creation of a European market for armaments.

The political breakthrough for the issue of armaments occurred when it was a focus of government attention in the defense field. The parallel discussion by the organizations in the market field was not sufficient to start any major European organizing activities. It was not until the defense field activated the question of a defense policy for the European Union that the issue of armaments became the object of political attention. However, when the governments activated the issue of armaments in the defense field, this codified a political and an industrial process that was already under way. Major changes have taken place in the European industrial landscape during the 1990s. National and state-owned companies—National champions—have been transformed into private European and transatlantic companies. This European consolidation process was especially salient in the mid–to late 1990s, but the process started earlier since the European restructurating process was preceded by changes in the national industrial landscape.

In what ways did the two fields approach each other? Did a common understanding evolve of how to frame and handle the issue of armaments, or did the organizations from the two fields only act in a way that was more coordinated than before? Hence, are we dealing with a situation in which various actors' interests from the market field and the defense field converged, or did they upgrade their common interest? The European Commission has been active in bringing the two frames closer to try to create a more comprehensive European policy and cooperation on armaments. A new understanding of the issue of armaments has emerged in the sense that the Commission, the working group in the Council (POLARM), and other organizations have expressed the view that the issue of armaments is complex and consists of both market and defense components. This is especially salient within the Commission, which has argued that the issue of armaments belongs to the first and second pillars of the European Union and that the regulative rules from these two pillars should be combined.

The rules of the game of organizational fields are seldom taken for granted but are contested and put into question. The crucial issue is whether the contested rules are regulative or constitutive. Regulative rules are formal rules, for instance, legal rules and decision-making procedures. Constitutive rules concern more fundamental ways of understanding and categorizing issues. Political tensions and discussions on regulative rules do not necessarily challenge the field boundaries. If, however, the discussions concern fundamental rules of the game—constitutive rules—the very basis of the field is put into question. The empirical analysis showed that the discussions of the organizations, for instance, those in the Commission and the Council, often concerned regulative rules, for instance, how to interpret Article 296 in TEU. The European Commission and EDIG, the European branch organization for the defense industry, wanted a stricter interpretation of the article. According to the Commission, some member states have interpreted the article broadly, which has led to the EU industry losing ground to the U.S. industry. An important argument for a stricter interpretation of the article was the perceived need for increased linkages between the civilian and defense-related spheres due to the changed dynamics in the technological and industrial sectors. Traditionally, it was the military sphere that gave the civilian sphere the technology—the so-called spin-off effect—but the spin-off effect has more or less been replaced by what has been dubbed the spin-in effect. This means that the defense industry is becoming more dependent on civilian industry and civilian RTD programs.

The Commission has therefore suggested that Article 296 should cover only sensitive goods. This position seems to lie near the view of the EU governments in the working group POLARM. It is, however, quite clear that the governments, especially the French, are in favor of an intergovernmental approach and they are very reluctant to strengthen the role of the Commission on issues of the defense industry. The reason for this position of POLARM is that certain prerogatives of national sovereignty should be maintained for security reasons but also that unconditional access to the European market could dangerously put pressure on Europe's defense technology industrial base from a strong American defense industry.

The underlying conflict over how to interpret Article 296 concerns constitutive rules—especially the question of whether the issue of armaments should be interpreted as belonging to the supranational and civilian pillar of the EU or to the intergovernmental and defense-oriented pillar. This is a classic conflict dimension of the EU. In this case, the conflict is highly problematic since all the actors involved showed an awareness of the problems with an issue that does not fit into the traditional organizational structure, that is, between the first and second pillars of the EU. The crucial question is thus not whether the issue is complex, but whether it implicates a reframing of, and a new way of organizing, the issue. The Commission has presented advanced suggestions for how the issue can be reframed and handled from both a market and a defense field. It has thus tried to change the traditional rules of the game and construct a new way of framing and organizing armaments. The work with the latest communication on the defense industry suggests that we are dealing with an informal horizontal and sectorial network within the Commission. We can identify a discrepancy between the pillar structure in terms of which the Commission is organized and the actual day-to-day work activities (Mörth 2000a). The Council did not pursue a new frame on the issue of armaments. It has instead discussed the issue from within a defense field. It is thus the EU's defense policy and its regulatory rules that will determine how the issue is framed and handled.

The empirical analysis showed that a single organization can belong to several fields at the same time. This was especially evident to the European Commission, which belongs to both a market field and a defense field when it comes to interpreting and conceptualizing the issue of armaments. The organizing activities within the Commission are based on both the first and second pillars of the EU. It is too simplistic an assumption in the literature on organizational fields that organizations belong to different fields. Organizations are often characterized by conflicting goals, heterogeneity, ambiguity, and unpredictability. This means that the boundaries within organizations seem to be as important as the boundaries between organizations. An empirical result is thus that tensions between the two fields and frames were obvious not only between the organizational fields but also *within* the fields. The two fields were thus not as coherent as the study set out in the theoretical chapter.

The contacts between the two organizational fields were intensified during the late 1990s and in the early parts of this century. Two European think tanks functioned as important nodes in a defense and industrial network in which representatives from the two organizational fields participated. However, the organizations did not always seem to be aware of each other's activities, which could be interpreted as ignorance or as part of an intensified organizing process.

OCCAR and LOI are difficult to classify in terms of the two the fields and frames—market and defense. They were primarily initiated and governed by an exclusive group of defense ministers and not by ministers dealing with traditional civilian issues. The two loose organizations emphasize market aspects, for instance, the importance of competition, commercial principles, and the principle of global *juste retour*. The political motive for establishing the organizations seems to be the defense industrial restructuring and that the transformation from national defense industrial companies to transnational and European companies needs political steering at the European level. The LOI governments emphasized in 1998 that they wanted to help the emerging transnational defense industrial companies. Initially, very little was mentioned regarding how new military tasks required a new way of cooperation on armaments. The initiatives show that the dichotomy between market and defense is regarded as problematic and that cross-pillar issues must be handled in a way that goes beyond the traditional boundaries between the defense and market fields.

The traditional and established organizations from the two fields—the EU, the WEU, WEAG, and WEAO—are challenged by the pretendents—OCCAR and LOI. The pressure is made possible due to the general political process after the end of the Cold War that has opened new possibilities for organizing cooperation on armaments. In contrast to the WEU, they are not organizations from the Cold War, and, in contrast to WEAG and WEAO, they are based on exclusive membership and on commercial principles. Counter to the Commission, the LOI governments are not convinced that the best way to cooperate and organize this issue lies in the European Union. In the LOI initiative the governments made it clear that the issue of armaments was an issue for the six governments and that it belonged primarily to the intergovernmental path of the European integration process. They did that by excluding any reference to the European Commission. Furthermore, by excluding any reference to POLARM, the LOI participants made it clear that LOI is something other than the European Union and that LOI is an actor in its own right. The LOI governments also emphasize the importance of mutual interdependence as the institutional glue between the participating countries. Thus, the lack of a hierarchical and supranational cooperative structure requires that the governments and the industry are organized and institutionalized through a looser and more informal way than would be the case if the cooperation had been based within the EU.

It can therefore be argued that OCCAR and LOI are important organizations in the emerging European organizational field on armaments. They represent new forms of domination and authority structures that are both formal and informal and based on the two frames and fields. OCCAR and LOI were initially very loose organizations based on voluntary agreements and network relations. Gradually these organizations have incorporated a more legal setup and have made legally binding agreements. The emphasis on the importance for an organization to have legal personality suggests that this is crucial to gain respect and to become an authoritative and legitimate actor. Paradoxically, the informal and flexible way of organizing is at the same time regarded as something positive and valuable. This is most clear in the reports from the LOI working groups, which quite often criticized the inflexibility of WEAG and WEAO. The latter are considered to be old-fashioned and in need of major changes. The empirical analysis showed that WEAG and WEAO have been active in pursuing a policy in which these two organizations will have a central role in a deepening European cooperation on armaments, for instance, in the work on the establishment of a European armaments agency. The political support for this work seems to be very weak due to the LOI initiative and the fact that the WEU is going to be incorporated into the EU. The future of WEAG and WEAO is therefore uncertain.

The actors in the study have used changes in the treaties and other important political decisions to give their frame authority and to legitimize a certain way of approaching the issue of armaments. The tensions over how to define the issue of armaments concern power and influence. It is also evident that the general political development has opened windows of opportunity for new types of organizing activities such as the LOI initiative. Actors have not only created room for maneuver but are also given such room (cf. Kingdon 1984). The LOI initiative meant that the governments were empowered vis-à-vis the European Commission and WEAG/ WEAO. In the formation of an organizational field on armaments some of the actors are empowered (LOI) and others seem to have lost power and influence in the emerging field (the Commission and WEAG). Furthermore, the frames not only have been instrumental for the actors involved, but they also seem to have functioned as important identity-building components. By interpreting events and monitoring other organizations, the actors communicate with other actors in their external environment and they try to find their own roles and places in the emerging field.

The empowerment of organizations during certain organizing periods suggests that different organizations can have various impacts and influences in the organizing process. Supranational organizations, such as the European Commission, play an important part during formative periods such as the agenda-setting process when the processes of institutionalizing and organizing are in their early stages, whereas the Council and the member governments play a crucial part at later stages of the consolidating process. In uncertain and turbulent political periods—in which a set of standard interpretations identifying the important problem and appropriate policy responses has not yet been formed—there is room for maneuver for various actors who want to influence the policy sector (cf. Keohane and Nye 1977). In the EU literature it is often argued that the influence of the European Commission is greatest in times of uncertainty, in which the member governments have no clear policy or common position and when the competences between the EU bodies are not yet defined (Cram 1997). "For under the given institutional conditions of messiness and lack of transparency, key actors, mostly the Commission, and the European Court of Justice . . . move into the vacuum created by shared and somewhat unclear structures of responsibility, and profit from the general uncertainty, overcoming the veto of actors' resistance by subterfuge" (Héritier 1999, 8-9). This study has shown that the Commission has played an important role in bringing the issue of armaments onto the political agenda and that the initiative was taken over by the governments within the LOI.

To sum up, the emerging European organizational field on armaments is weak in terms of sociological institutionalism's definition of institutions. The constitutive rules are contested. However, the empirical analysis shows that there is a process of institutionalization that entails the issue of armaments being reframed and interpreted as an issue that goes beyond the dichotomy between the defense and market fields. In addition, new authority and power relations have emerged, and new patterns of inter- action and communication between and within organizations have been identified. This is a field in the making and its autonomy is therefore weak. We can thus describe the organizational field as thin. This means that the two fields have moved closer in terms of overlapping activities and issues, that is, they both deal with the issue of armaments and therefore have to interact with each other. The empirical evidence suggests that it will gra- dually become thick in terms of institutionalized rules, frames, identities, organizational authority, and power relations among different organiza- tions from the two fields. The study suggests that the end of the Cold War has opened possibilities for new types of frames, organizations, and orga- nizing activities which in a more fundamental way change the rules of the game for European cooperation on this issue.

The organizations have not reached a phase that can be described as institutional isomorphism. The study shows, however, that one important driving force behind field changes—from two separate fields and activi- ties into a situation that is characterized by diffuse field boundaries and an emerging field—is mimetic. The chapter on the market field showed that the American restructuring of its defense industry has been an important rationale behind the Commission's and the industry's work related to put- ting the issue of armaments on the European political agenda and to esta- blishing European cooperation on armaments. The LOI initiative was also driven by the notion of how the Europeans are lagging behind the United States. The industrial restructuring in Europe clearly resembles the chan- ges in the industrial landscape in the United States—from multiple natio- nal defense companies to only two or three large transnational entities.

This study has shown that different fields produce different types of knowledge and images of the world. To present different authoritative ver- sions of valid problems and relevant solutions to the problems is important in every political activity. The actors in the study—especially the Com- mission and the intergovernmental organizations—pursue different frames depending on their interests, background, and positions in the process, that is, depending on whether they belong to the European political economy project or the security and defense project. The study has also shown that organizations influence and change the rules of the game. During turbu- lent political periods, windows of opportunity create room for change. Another general conclusion from the study is that organizing European cooperation evolves slowly and in a piecemeal fashion. Sometimes there is a rather rapid political development that creates room for reorganizing. The reorganizing—the formation of a new European organizational field —intersects with preexisting fields in which there are long–standing ways of doing things and of defining issues (cf. Green Cowles et al. 2001).

Drawing on the empirical study on the formation of a European organizational field on armaments, in the following sections of the chapter I discuss three themes of European politics. The first theme concerns flexible integration. The recent creation of restricted clubs between a few industrially strong European countries, OCCAR, and the LOI agreement raises the question of flexible integration. How politically acceptable are flexible arrangements in European armaments cooperation? This is a general question in the EU and European politics. It concerns the very heart of European cooperation, namely, how to accommodate both diversity and unity.

The second theme concerns another fundamental question that is raised in the empirical analysis—that of the relationship between Europe and the United States. The major driving force behind the activity on the issue of armaments has been the image that Europe is lagging behind the United States. The Commission, industry, the LOI governments, etc., have had the ambition of creating a European policy and a European actor vis-à-vis the United States. Clearly, the United States is important in Europe's identity-building process, but is it a foe or a friend?

The third and concluding theme concerns boundaries and European governance. This study on organizational fields concerns boundaries of many kinds. One important border is that between issues/actors that are considered to be within the field in question and issues/actors that are considered to exist outside the field. How we define boundaries between political issues and how we frame issues are important in EU policymaking. To frame the Economic and Monetary Union (EMU) project as a politically driven project or as an economically driven project has important power implications since the two frames activate different sets of problems, actors, and rules. Power over the political agenda is thus important in the policymaking process. This is a well-established notion in political science. A less elaborated question among students of the EU and European politics is how we should analyze and study frame competition and issues that cross organizational boundaries and do not fit into our traditional ways of studying European politics. When traditional policy boundaries are questioned, new rules of the game emerges. These complex changes of European governance are not often easy to study empirically if we want to keep an analytical focus. The concluding part of the chapter will therefore discuss how we can analyze the complexities of European governance without losing analytical clarity in a totally different empirical case with the use of an organizational field approach—a European organizational field of human rights.

LOI and Flexible Integration

It is important to have a comparative policy perspective when analyzing European cooperation. Students of CFSP and the first pillar of the European Union seldom draw any empirical or analytical conclusions from each other but tend to discuss only empirical cases in comparison with other case studies within the relevant EU pillar. This is unfortunate since there are interesting similarities between apparently different policy processes. One similarity is, of course, the fact that political deadlocks are common in the EU as a whole. The explanations for the separate stalemates are multiple. The interesting point is, however, that change does occur and the question is therefore whether there are any general mechanisms behind changes within different policy processes (Héritier 1999). This is not an easy question to answer since an answer would have to be found by way of thorough empirical studies from many policy processes. My ambition in this section of the book is, however, to discuss parts of my study of European organizing on armaments in a broader empirical context.

One general conclusion that can be drawn from reading about the EU and European integration is that political conflicts and other problems have not prevented those governments that wanted to develop cooperation further from doing so. Indeed, "much of the history of the EU has been about efforts to find formulas, institutions and policy regimes which weave forms of unity out of diversity" (Wallace 2000, 185). A classic problem in the EU is the matter of how internal cohesion and mutual solidarity can be upheld at the same time as a limited number of states wish to deepen the integration process within a policy area. This tension is considered to be the strongest that it has ever been in the EU's history (Stubb 2000). The Schengen agreement, the Social Protocol, and EMU are all examples of flexible integration. Flexible integration is a complex principle since it entails different forms of differentiation and decision-making rules. As a general principle it means that some member states cooperate without the others and that EU membership is differentiated. Although this principle is becoming a more common feature in the EU, it goes against the fundamental principle of the EU that community method assumes full membership (Jerneck 2001).

European organizing and cooperation are complex and multifaceted. The reasons behind this mosaic of European cooperation and organizing are many and varied. One important dynamic behind flexible integration and differentiated membership is political stalemates and the resulting inertia in the EU institutions. This creates a pressure for activities by a smaller (and exclusive) group of countries to organize and deepen their policy cooperation. We can identify three different models of flexibility. The first one is Europe with multiple speeds, the second is Europe à la carte, and the third is variable geometry (Stubb 2000; Jerneck 2001). The first model entails a commitment among the EU member states to achieve the same goal but at different speeds. The second model means that the member states can choose from the EU menu according to their own interests and needs. In the third model it is assumed that the member states "take part in a number of common policy areas while the possibility remains open for a group of states to take a further step with the purpose of deepening the cooperation" (Jerneck 2001, 157). Thus, the institutional structure is common for all member states, but the model allows for a group of states to deepen their cooperation.

Flexible integration has been a permanent feature since the early days of the European integration process. The general political view of the principle has changed. Differentiated terms for member states began as necessary solutions in difficult situations. Flexible clauses have gradually become a legitimate way of handling diversity of interests and conflicts over the scope and depth of the EU. The political and judicial breakthrough came in the negotiations over the Maastricht Treaty when the British government refused to sign the Social Charter and the monetary cooperation agreement (Jerneck 2001, 157). In 1995 the French and the German political leaders called for a treaty-based principle that would enable certain EU member governments to coordinate its politics. Flexible integration was formally incorporated as a general principle in the Amsterdam Treaty (Stubb 2000). The treaty does not mention flexible integration; the term used is "closer co–operation" (Article 11 and Articles 34-45). At the Nice Summit, member states seem to have agreed in principle on the idea of removing the emergency brake on enhanced cooperation (which is the new term for flexible integration) and allowing groups of member states to forge ahead without all of them having to be involved. This suggests that flexible or enhanced cooperation will be a permanent and an important principle in the EU.

Examples of flexible integration show that these have been gradually incorporated into the existing EU political and legal framework. This is clearly the case with the Schengen agreement. The Schengen agreement and the Social Charter were, in 1997, incorporated into the Amsterdam Treaty. The case of EMU differs from both the Schengen agreement and the Social Charter in the sense that it was initially already part of the EU. In the case of the Social Charter it was decided that the eleven countries should enter into a Social Protocol that was linked to the Maastricht Treaty (Bernitz 2001). In contrast to the Schengen agreement, the Social Protocol was signed by all EU member states except the United Kingdom. In the Schengen case, the escape route (Héritier 1997) from the political deadlock in the European Union comprised only a core of states. The Schengen agreement on the abolition of internal border controls started as a Franco—German agreement and was signed in 1985 by France, Germany, and the three Benelux countries.

The recent LOI agreement on the defense industry and equipment shows several similarities with the case of Schengen. During many years, until the mid-1980s, the national resistance to the removal of internal frontier controls was high (Den Boer and Wallace 2000). The visibility of the Schengen agreement was directly linked "to anxieties about statehood.... particularly in France, which flowed from the fears articulated by politicians and journalists that France without frontiers would have lost control of the national territory" (Den Boer and Wallace 2000, 503). However, concrete problems with the cross-border traffic decreased the national fear of losing state sovereignty over its territory. A comprehensive transbureacratic cooperation and interaction between various national agencies seemed also to have increased the trust among the governments and paved the way for the removal of national barriers. The tension between integration and state control over their internal borders was, however, evident in the implementation phase of the agreement. When the rhetoric on the removal of barriers between countries became implemented in practice, intergovernmental mutual trust and confidence was put to the test. This proved to be difficult and the mutual trust concerning how states controlled their external borders was therefore reinforced "by an intergovernmental system of inspection, in which teams of officials from member governments checked on the adequacy of controls on the external borders" (Den Boer and Wallace 2000).

As in the case of the Schengen agreement, the national resistance to removing national control of the defense industry has been immense. This is easily illustrated by the discussion of Article 296 in the TEU and the political deadlock in the Council (POLARM) on the communications from the Commission. Control over the national territory and the control over the national defense industry clearly concern core areas of national sovereignty. However, cross-border activities by traffic or by the defense industry reduce the state's ability to control the policy areas. Instead, they seek cooperation at the European political level. The governments are thus pressured to seek ways of handling traditionally domestic policies. They take part in various European cooperative arrangements that cannot be reduced to an aggregation of the various national interests. LOI is an actor in its own right in the sense that the governments collectively reach a mutual understanding of how to deal with the European defense industry and its implications for national procurement policies, exports of armaments, etc. They make joint decisions and shift their expectations to this new form of organization. The organizational setup of LOI showed that the ministries and the state agencies from the six countries were heavily involved in the work. States—the political leadership and their bureaucracies—are not outside of the European organizing process. They are part of a wider European organizational and institutional context. In addition, the governments within the LOI initiative and the framework agreement have committed themselves in such a way that it is not relevant to speak in terms of *inter*governmental cooperation.

Schengen, the European military cooperation with the WEU, and the LOI show that cooperation between EU member states can formally lie outside the EU. Indeed, some cooperative arrangements lie within the core of the EU and the integration process whereas others are linked more loosely to the EU. Thus, the LOI initiative and the framework agreement do not fit into any of those models of flexible integration since they formally lie outside the EU. It is, however, important to note that LOI is not taking place in an organizational and institutional vacuum. The LOI initiative and the framework agreement will legally bind only the six governments. The agreements are, however, dependent on and embedded in a larger European political and regulatory environment. The political considerations that have to be taken are multiple. How will the governments act toward WEAG? WEAG will probably continue to exist since it is the only pan-European organization in which the European countries (not only the EU member states) can discuss the issue of armaments. The relationship to the EU is even more complex. The LOI initiative and the framework agreement were a way for the six governments to take a lead in the organizing process and bypass the Commission, POLARM, and WEAG. It can also be argued that the development within the LOI would not have been possible without the EU and its regulatory rules. The framework agreement explicitly mentions the EU's code of conduct for arms in connection to the issue of transfer and export procedures with countries outside the LOI circle. This has been interpreted as the important legitimation base for the entire LOI agreement. The six governments thereby guarantee that they will not pursue a different export policy from what is decided within the EU.

Europe and the United States—Friends or Foes?

A major driving force in the formation of a European organizational field on armaments is the European relationship with its external environment —European organizing cannot be analyzed as a self-contained entity. To understand the European organizational activity we must analyze Europe's relationship with the United States. As in other constructions of collective identity, a definition of the Other is needed to construct a European actor and identity (Stråth 2000). The basic idea in this line of thinking is that the organized structure of identities and rules is not static but changes in response to external and internal pressures (March 1994). An important dynamic behind the creation of the internal market was the perceived economic threat that the Strategic Defense Initiative posed to the Europeans. Nineteen ninety-two "is a vision as much as a program—a vision of Europe's place in the world. . . . We propose that changes in the distribution of economic power triggered the 1992 process . . . shifts in relative technological, industrial, and economic capabilities are forcing Europeans to rethink their economic goals and interests as well as the means appropriate for achieving them" (Sandholtz et al. 1992, 82). This conclusion fits well into a general explanation of the increasing focus in European politics on European identity. "European identity was put on the agenda just as political economy, in its Keynesian form that had been established since the 1950s, was becoming exposed to severe strains" (Stråth 2000, 401).[1] A crisis for national economic governance led to a new European political economy paradigm that highlighted concepts such as market, flexibility, and deregulation.

In the case of armaments, the Commission, the defense industry, policy centers, and media have been active in constructing a market frame to legitimate various European responses, for instance, to deregulate the national defense industry and to communitarize the issue of armaments. This frame can be characterized as a threat image. A threat is not only instrumental for actors in their pursuit of interests; it is also important in the European identity-building processes. The emergence of collective identities is dependent on the creation of boundaries, which is in turn enhanced by a threat image (cf. Cederman 2001; see also Eriksson 2001). The technological and economic market frame on the issue of armaments organizes collective experience and gives meaning to events and occurrences. The suggestions from the Commission, the LOI initiative, etc., are part of a process in which a European actor is created. The creation of binding agreements between the six countries on how to establish centers of excellence on the production of armaments across the countries is perceived as important in order to stand against the United States.

The American administration and the European governments have a long and complex history of rivalry over issues that concern information technology and other strategic and defense–related technologies (Mastanduno 1992; Mörth and Sundelius 1993). The administration in the United States has traditionally handled these issues bilaterally with the European governments. In February 2000 a Declaration of Principles for Defense Equipment and Industrial Cooperation was signed between the United Kingdom and the United States (Axelson and James 2000). The American administration is also discussing such declarations with France and Germany (Axelson and James 2000). The LOI agreement requires that the European governments create an actor, a buffer in relation to the United States, and that negotiations cannot be handled unilaterally but only by way of the LOI countries as a collective. This decreases the chances of outside pressures. There are also indications that the United States is easing its pressure.

The case of armaments shows that the United States seems to be a friend and a foe at the same time. It is also quite clear that the fear of the United States resonates differently in the different European countries and within their political leaderships.[2] For the European defense industry, the relationship seems less tense since it acts according to the logic of the market rather than according to more political and security-related considerations. In chapter 4 it was shown that the United States has been a model for European defense industrial restructuring and that the big European aerospace companies are seeking close relationships with American companies.

Scholars who study the linkage between security and international political economy suggest that the end of the Cold War has led to a power struggle between states over economic and technological terms. States' struggle for economic growth, knowledge, and competitiveness has partly replaced the more military and territorially oriented security policy (Crawford 1995, 158; Strange 1992). Nye argues that "[i]n assessing power in the information age, the importance of technology, education, and institutional flexibility has risen whereas that of geography, population and raw materials has fallen" (1996, 22). The popular notion of competitiveness "has been widely used to create a sense that the US and Europe are 'losing' to each other in some kind of knock-out competition" (Cable 1995, 310; Hart 1992; Tyson 1992).[3] The increasing competitiveness is thus often described in realist terms—as economic warfare between leading countries in the world. There is a struggle between independent actors striving to maximize their own utility—the classic logic of anarchy in the international system. The United States, the EU, and Japan (and increasingly China) "are essentially adversaries though the weapons in countering threats to national security are economic policy measures rather than Cruise missiles and Stealth bombers. By combining a 'realist', Machiavellian, approach to international relations with the language of security and the economic insights of 'strategic trade theory', advocates of a more mercantilist approach have achieved some intellectual respectability and made some impact, in the US especially" (Cable 1995, 307).

The rhetoric within the market-based EU not only conveys the message that the European countries are interdependent, but also emphasizes the notion of a threat from the United States. In NATO and in the defense-based EU, the message is that there is increased mutual interdependence between countries. Hence, the traditional rhetoric within the sphere of interdependence is moving into defense-based organizations, whereas traditional anarchic rhetoric is becoming a common feature in the market-based organizations.

Boundaries and European Governance

This book has studied how European organizing has emerged around an issue that crosses the boundary between the European political economy project and the defense and security–oriented project. We are dealing with parallel and nested organizational activities. As Helen Wallace puts it, "the development of the EC into the EU has raised expectations that two different projects of integration might be elided, namely, the political economy project, developed through the EC, and the defense and security project, hitherto organized through NATO and the WEU" (Wallace 2000, 177).

A general understanding among students of the EU and the European integration process is that we have rather a deep knowledge of various policy areas. "A rich set of case studies now exists on the formulation and implementation processes of Community policies" (Lequesne 2000, 42; see also Wallace and Wallace 2000) But what kind of knowledge do we have on policy areas that in one way or another are linked to each other?

Cross-pillar studies require a research strategy that uses concepts and an analytical framework that allow a broad empirical investigation. The purpose of such studies is to obtain a picture of the political activities that is as comprehensive as possible. In this book it is argued that the organizational field concept allows for a more flexible approach than the traditional approaches on European integration. It is not necessary to decide beforehand to which European project the issue of armaments belongs. It belongs to both, and it therefore activates the supranational and intergovernmental parts of the EU, the WEU, NATO, and new types of organizing activities. Thus, it is necessary not only to go beyond the traditional dichotomy between the civilian and defense orientation, but also to widen the study to include organizational actors that are only loosely coupled to the EU. Thus, the strength of an organizational approach to European integration is that it simultaneously focuses on multiple and parallel processes and actors that participate and influence each other. Furthermore, it also focuses on how the interactions—between the organizations—evolve through a process of institutionalization.

The empirical facts in the study are in themselves rather well known. Students of European integration have for a long time studied the EU's market-making activities and the process toward a common foreign and security policy. However, they have analyzed these two activities as separate processes. My study has shown that these two traits in European political studies must be combined and that a combination of the two processes gives a richer and more multifaceted picture of the empirical complexity of European politics than would have been the case if the issue of armaments had been analyzed either as part of the EU's RTD/industrial policies or as part of the EU's emerging defense policy. In so doing, this study has uncovered deeper processes within the European polity and has shown how these can be studied with a new type of analysis of European integration.

The end of the Cold War has clearly meant a rather turbulent organizational period in which the question of how to frame issues is more contested than ever. New economic and security challenges have emerged and a new policy paradigm has evolved that entails RTD and industrial issues being closely linked to the EU's defense capacity. A change of policy means that issues of authority are central. "In other words, the movement from one paradigm to another is likely to be preceded by significant shifts in the locus of authority over policy" (Hall 1993, 280). New authority relations are constructed that institutionalize a policy within which both defense and markets matter. This is what the field analysis showed. Consequently, questions such as which organization an issue belongs to, and what kind of problems, rules, etc., are relevant, are more open. European organizing has become more complex and multifaceted than during the period of the Cold War. This means that students of the EU and European politics should go beyond the traditional intradisciplinary boundaries. What had traditionally been an acceptable simplification of European politics has today become obsolete and therefore an unacceptable approach in the search for the dynamics in European politics.

What type of analysis is feasible if we want to study empirical processes that belong to different analytical traditions in the study of European politics? In what ways has this case study helped us in that endeavor? How we study European politics seems to be a result of tradition and habit. The early students of European integration approached the EC from an international relations perspective and focused on the driving forces behind the integration process. The studies on the EU and European integration have since broadened along with the deepening of political and economic integration. The political and economic development in the EU has also made theories and concepts from various subdisciplines in political science relevant (Rosamond 2000; see also Hix 1994; 1996). The EU has evolved into a mature polity that should be analyzed by the traditional political analytical questions of who gets what and when. The governance turn in the study of European integration has opened possibilities for new analytical and theoretical perspectives. New theoretical combinations have emerged, for instance, between neoinstitutionalism and the study of European integration (Aspinwall and Schneider 2000). In my analysis of a cross–empirical case I have used insights and concepts from both the general literature on politics (the concepts of regulative and constitutive rules and frame) and the organizational literature (the concepts of organizations, organizing, and organizational fields) to identify the complexity of European politics with respect to armaments. I combined these concepts and theoretical approaches to avoid the simplistic dichotomy between an intergovernmental and supranational integration process. The literature on multilevel governance is not (yet) theoretically elaborated and is therefore less useful in the analysis on how new European authority structures evolve. However, my study fits well into its basic standpoint that the boundaries between levels of government are diffuse and that many types of actors are active in European organizing (Hooghe and Marks 2001).

The literature on organizations/organizing and on multilevel governance seems to share the idea that authority relations not only are backed by a hierarchical and legislative form of authorization and legitimization, but also that there are looser structures based on networks and voluntary agreements. A general notion among students of the European Union is that there is a growing preference for noncoercive and non-legally binding governance in the European Union, that is, soft law. Majone argues that this shift from command and control to modes of regulation based on information and persuasion "should be seen as part of a general reappraisal of the role of public policy in an increasingly complex and interdependent world" (Majone 1997, 269). What is interesting with this notion of the increased noncoercive rules within the EU is that the authority and legitimacy must be based on something other than the traditional Weberian rational—legal authority (Boli and Thomas 1999). Recent studies of the EU suggest that the political arena "is populated by formally autonomous actors who are linked by multifaceted interdependencies" (Eising and Kohler-Koch 1999, 4). The EU is characterized by network governance that gives order and patterns to the cooperation despite the lack of any central organizing authority. "The concern with instrumental action, means-ends rationality, effectiveness and efficiency in reaching predetermined goals has been supplemented with an interest in communicative behaviour and discourses on values, ethics and morals, as a necessary condition for

explaining cooperation, civilized conflict resolution and order" (Brunsson and Olsen 1998, 17). Hence, while government is based on hierarchy,
monopoly, and democratic legitimacy, governance is based on networks,
competition, and knowledge. The use of soft law is thus part of new modes
of governance in Europe (Héritier 2001), which accounts for the conclusion that governments have lost some of their authority in the policy-making process and that they have to share it with supranational and private
actors (Börzel and Risse 2000).

The notion of informal organizing activities and noncoercive rules is
crucial not only in recent thinking on the European Union. The phenom-
enon of governance without government lies at the core of the study of
international relations. This literature offers a way of looking at international authority that does not require formal super/subordinate relations. The
regime literature helps us to explain how mutual obligations may emerge
without the structure of those authority relations conforming to traditional
expectations (Ruggie 1998a; 1998b). Hence, the lack of formal agreements and hierarchical relations and formal transfer of sovereignty within
a policy area cannot automatically be interpreted as a weak process of institutionalization (cf. Brunsson 1999). International regimes consist of formal and informal rules, which are "implicit or explicit principles, norms,
rules, and decision-making procedures around which actor expectations
converge in a given area of international relations" (see Krasner 1983, 1).
The regime literature is thus focused on how new authority relations are
created. It is assumed that consensual ideas, that is, shared causal beliefs,
will emerge (Haas 1992; Hall 1993). The concept of organizational field
is also focused on how new authority relations emerge, but also on how
organizations compete with each other. The process of field consolidation
does not necessarily lead to a situation characterized by harmony and lack
of political conflicts.[4] Institutions are seldom uncontested. They can more
or less be taken for granted. We should therefore study processes of institutionalization rather than institutions themselves.

To sum up, the formation of European cooperation on armaments has,
in my study, been analyzed as an organizing process in which institutionalized organizations interact. European cooperation has thus been analyzed as a process that gradually takes shape in terms of the organizational
setup but also in terms of institutions, that is, rules that guide collective
behavior. The concept that was used to analyze the organizing process
was organizational fields. This concept enabled me to broaden the study
empirically. Studying European integration as organizing permitted an in-

clusion in the analysis of all organizations and organizing activities operating in the same domain that critically influence each other's performance, irrespective of whether they are formal/informal organizations inside or outside the formal structure of the EU. Organizational fields consist of organizations that are held together by institutionalized rules. These rules determine how issues are interpreted and categorized. Indeed, how issues are framed and put into a certain political context is determined by institutionalized rules of the organizational fields. These rules can be conceptualized as cognitive borders of the external environment. To paraphrase the expression, Where you sit is where you stand—"Tell me what your borders are and I will tell you who you are" (Zielonka 2002, 7).

The organizational field approach that has been used in this study can be used in other empirical case studies. One case in point is the process toward a European organizational field on human rights.

The division between foreign policy and other externally oriented policy areas depends on the existence of borders between internal and external policies. During recent years, however, the borders between internal and external policies have become more difficult than ever to uphold. The ambition to create a political community has above all meant that the EU seeks to develop policy in all relevant areas. Human rights, democracy and the rule of law are values that are important in EU policy areas such as foreign trade and aid. According to TEU, the CFSP (now European Security and Defense Policy [ESDP]) "will help 'to develop and consolidate democracy and the rule of law, and respect for human rights and fundamental freedoms" (Ehrhart 2002, 26). Indeed, it could be argued that a value-based external policy has emerged in Europe (cf. Haaland Matláry 2002). A host of "new issues have emerged that are truly common to many states: environmental problems, conflict resolution and peace implementation, terrorism and so on. . . . Poverty, immigration, the resource depletion of the seas—these are all examples of the new type of issues where we can often not isolate any national interest" (Haaland Matláry 2002, 25). These issues cannot be dealt with through the different political and legal clear-cut definitions of the pillar structure. The process toward a European value-based foreign policy means that the borders to other policy areas in the first and third pillars are blurred. Foreign policy includes trade, aid and development, justice and home affairs, environment, and sustainability. This has led the European Commission to argue for a merger of the three pillars (COM (2002) 247 final). This could give the EU a better base for coherent and effective policy making in international politics. The range of instruments and decisionmaking procedures will however be retained according to the Commission (COM (2002) 247 final). The diverse legal basis in the pillars is of course one difficulty if the aim is to create a better actor capability for the EU on foreign policy. This is not only a legal issue. It is above all a political issue since the different legal bases of the pillars give varying roles and powers to the Commission, the Council, the European Parliament, and the European Court of Justice. The Commission has a strong role in trade and aid that can be used to support and implement wider foreign policy objectives, whereas the Council determines the ESDP framework. The economic instruments such as embargo and conclusion of trade agreements generally fall under the European Community's jurisdiction (Smith 1998). These instruments are increasingly used to support the policy making in the second pillar. This is the case with the use of economic instruments (conditionality) to "encourage democratic reforms and respect for human rights" (Smith 1998, 73) that have become an important part of EU's foreign policy.

Human rights and other values are pursued by Western governments and regional and global organizations. The number of international arenas for norm creation and norms export has increased (Haaland Matláry 2002). This is especially salient in the UN family of organizations but also in the OSCE (Organization for Security and Cooperation in Europe), COE (Council of Europe), the WTO, OECD, and EU. NATO has also strengthened its emphasis on the linkage between security and democracy, human rights, and the rule of law. These organizations are based on different legal (regulative) and constitutive (the fundamental aims) rules. The Council of Europe and the OSCE have traditionally been strong organizations in issues of various norms—soft power (Nye and Owens 1996). The OSCE's human rights instruments are politically rather than legally binding. The European Convention on Human Rights (ECHR), a legally binding convention, entered into force in 1953, whereas other conventions rely on weaker enforcement mechanisms (Haaland Matláry 2002). The EU's political interest in human rights issues can be dated to the Single European Act of 1986, when the concept is mentioned in passing (Haaland Matláry 2002). It is, however, not until in the Treaty on European Union, that it is stated that the EU is "founded on the principles of liberty, democracy, respect for human rights and fundamental freedoms, and the rule of law" (Haaland Matláry 2002, 184). A suspension clause in the case of breaches of human rights commitments was included in the amendments in the Amsterdam Treaty, and the requirement that the candidate countries must respect the principles if they are to become EU members (Haaland Matláry 2002; see also the human rights clause, adopted at the Copenhagen European Council in 1993).

There are also a growing number of various nongovernmental organizations that form different value-based advocacy networks. Furthermore, the notion of corporate social responsibility, the Global Compact suggest that multinational corporations are taking on roles, otherwise handled by state actors, on how to behave in relation to human rights, environment, and employment standards. New organizations have also embedded the values, "such as new institutions like the tribunals for Yugoslavia and Rwanda as well as the International Criminal Court (ICC)" (Haaland Matláry 2002, 10). The increased legalization of world politics—the juridification of human rights—changes the rules of the game in the sense that traditional state sovereignty and control are weakened. It is a situation in which different organizations—formal as well as informal—compete with each other on how to pursue values such as human rights, democracy, and rule of law. The organizations have different origins, memberships, legal and political instruments, and mandates. They are also interdependent, since the work on human rights etc., permeates many policy areas that cross geographical and political levels. This is tricky since the roles and identities of formal international organizations are achieved by getting international attention and recognition. The rivalry between the OSCE and the UN in Kosovo is well known (Haaland Matláry 2002).

The establishment of European cooperation on human rights and values can be analyzed as an organizing process in which institutionalized organizations interact with each other. An organizational field approach enables us to include in the analysis all organizations and organizing activities that operate in the same domain irrespective of whether the organizations are states, the EU, the OSCE, or private organizations. We can then study the organizations' institutionalized rules, and also how their activities create new institutionalized rules (regulative and constitutive) on the issue of human rights.

In the consolidating process of the formation of a European organizational field on human rights and other values, human rights and democracy will be conceptualized and defined. On the surface there seems to be a consensus on what the concepts stand for, but I am sure that an analysis on the issue of framing will reveal frame competition. The framing process will also include the reformulation of the distinction between foreign policy on the one hand and traditional internal policy areas on the other hand. The process toward an organizational field on human rights also entails new forms of formal/informal domination and authority structures, that is, the legalization of human rights, the EU's rather new role as an exporter of values, etc. EU's Charter of Fundamental Rights, and the discussion of whether it should be legally binding, and its relationship with the European Convention on Human Rights suggest that we are dealing with a discussion of who owns the question of human rights. This discussion also suggests that there is a tension between organizations that have a long and established role in dealing with value-based politics and organizations that have had a more market-based identity.

It is of course an empirical question if the organizations that are dealing with human rights issues also interact with each other, and if so, how the interactions can be described in scope and depth. It is also an empirical question if this emerging organizational field is thin or thick in terms of regulative and constitutive rules. Are we dealing with overlaps between different organizations such as the OSCE, Council of Europe (COE), NATO, and the EU that force the actors to establish closer cooperation on human rights? Or does the formation of European cooperation on human rights illustrate more fundamental changes in the conceptualization and handling of human rights, democracy, and rule of law?

The complexities of organizational fields suggest that the EU resembles a neomedieval empire rather than the traditional Westphalian state (Zielonka 2001). The latter is about "concentration of power, hierarchy, sovereignty and clear-cut identity, whereas the former is about overlapping authorities, divided sovereignty, diversified institutional arrangements and multiple identities" (Zielonka 2001, 509). It is my belief that an organizational approach, especially a sociological institutional way of thinking about organizations, is well suited for an EU that is characterized by important nonhierarchical networks and partnerships of public and private actors who increasingly choose to regulate politics by loose legal instruments (cf Marks and Hooghe 2001). Indeed, concepts of institutionalized organizations and organizing process identify the complexities of European governance without losing analytical clarity.

Notes

1. Stråth argues that the idea of a European identity was launched in 1973 at

the European Community Summit in Copenhagen in a situation characterized by a "growing lack of feelings of community and identity" (Stråth 2000, 403).

2. A common understanding among students of European integration is that the British political elite considers Europe to be a friendly Other, "whereas German elites have seen their country's own catastrophic past as 'the Other', while French political elites have traditionally added the US to their list of 'Others'" (Marcussen et al. 1999, 329).

3. See also Thurow 1992; Magaziner and Patinkin 1990; Garten 1992.

4. In the international relations-oriented literature on complex interdependence the existence of conflicts is seldom discussed. Mutual interdependencies between states are instead interpreted as a lack of conflict. Professor Bengt Sundelius has, however, argued that complex interdependence does not exclude conflicts. He has therefore introduced the term *adversarial interdependence* (1989).

Appendix: Interviews

af Sillén, Jan. (Ambassador, Swedish Ministry of Industry and Trade Ministry for industrial affairs). 1998. Interview by author. February. Brussels.

Ahlén, Gunnar. (Scientific Research Counselor, Permanent Swedish representation). 1996. Interview by author. December. Brussels.

——. 1997. Interview by author. February. Brussels.

——. 1997. Interview by author. June . Brussels.

——. 1997. Interview by author. September. Brussels

Åkesson, Ingvar. (Swedish Ministry of Defence). Interview by author. August. Stockholm.

André, Dag. (Scientific Research Counselor, Permanent Swedish representation). 1998. Interview by author. May. Brussels.

Andrén, Krister. (Swedish National Armaments Director Representative in CNAD and WEAG).1997. Interview by author. June. Brussels.

——. 1998. Interview by author. February. Brussels.

—— 1999. Interview by author. September. Brussels.

Beckman, Eva. (Swedish Ministry of Industry and Trade). 1997. Interview by author. January. Brussels.

——. 1998. Interview by author. April. Stockholm.

Beskow, Charlotte. (European Commission, Telecommunications, Information Market & Exploitation of Research (DG XIII). 1997. Interview by author. January. Brussels.

de Boissezon, Birgit. (Expert, European Commission, Science, Research & Development (DG XII)). 1997. Interview by author. September. Brussels.

Buffeteau, Francois. (Conseil économique de la Défence, Ministère de la Défence). 2000. Interview by author. Paris. March.

Carlsson, Gunilla. (Swedish Member of European Parliament [MEP]. Chairman of ADRIANE and member of Committee on Economic and Monetary Affairs and Industrial Policy). 1999. Interview by author. September. Brussels.

Cars, Hadar. (Swedish MEP. Committee on Foreign Affairs, Security and Defense Policy). 1997. Interview by author. September. Brussels.

———. 1999. Interview by author. January. Stockholm

———. 1999. Interview by author. September. Brussels

Chevallard, Giancarlo. (Head of Unit, European Commission [DG IA]). 1997. Interview by author. February. Brussels.

———. 1999. Interview by author. September. Brussels.

de Chantériac Arnaud. (Advisor to the Chairman of International Affairs, Défence Conseil International). 2000. Interview by author. March. Paris.

Delhotte, Pierre. (Head, WEAG Armaments Secretariat). 1998. Interview by author. May. Brussels.

———. 1999. Interview by author. September. Brussels.

Dumas, Pierre. (WRC General Manager, WEAG/WEAO). 1999. Interview by author. September. Brussels.

Engdahl-Ohlsson, Eva-Maria. (European Commission, Science, Research & Development [DG XII]). 1997. Interview by author. January. Brussels.

Eriksson, Anders. (National Defence Research Establishment). 1997. Interview by author. Stockholm. September.

Fagerberg, Lars. (Colonel, Swedish Ministry of Defence [Swedish Member of the Executive Committee of LOI]). 1999. Interview by author. August Stockholm.

Finocchio, Petro. (Major General, Panel II/RTC Chairman, WEAG/WEAO). 1999. Interview by author. September. Brussels.

Frey, Florence. (Expert, European Commission, External Political Relations DG IA]). Interview by author. May. Brussels.

Frier, Laurie. (DAS, Ministère de la Défence). Interview by author. March. Paris.

Gunnarsson, Pierre. (European Commission, Industrial Affairs [DG III]). 1998. Interview by author. May. Brussels.

———. 1999. Interview by author. September. Brussels.

Hessulf, Vulff. (Försvarets Materielverk [FMV], Swedish expert in the Master-plan for the European Armaments Agency). 1998. Interview by author. February. Stockholm.

———.1999. Interview by author. August. Stockholm.

Holm, Ulf. (Swedish MEP. Committee on Research, Technological Development, and Energy). 1997. Interview by author. January. Brussels.

———. 1998. Interview by author. May. Brussels.

Holmberg, Marie. (Swedish PFP representative, NATO). Interview by author. September. Brussels.

Huusela, Piia. (Information Officer, European Commission, DG XII), 1997. Interview by author. June. Brussels.

———. 1997. Interview by author. September. Brussels.

———. 1999. Interview by author. September. Brussels.

Ivarsson, Tommy. (Senior Vice President, Head of Corporate Strategic Planning, Saab AB). 1998. Interview by author. February. Linköping.

Jentzen, Herman. (Swedish Ministry of Defence [Swedish representative in POLARM]). 1998. Interview by author. February. Stockholm.

———. 1998, Interview by author. December. Stockholm.

———. 1999. Interview by author. August. Stockholm.

Jonsson, Olle. (Swedish representative from Ministry of Defence/Swedish Agency for Civil Emergency Planning, NATO). 1999. Interview by author. May. Brussels.

Karlström, Håkan. (Energy policy Counselor, Permanent Swedish Representation). 1997. Interview by author. January. Brussels.

Kirtikumar, Mehta. (Head of Unit, European Commission, Competition [DG IV]). 1999. Interview by author. September. Brussels.

Köhli, Penttii. (Senior Vice President Technology, Celsius Corp.). 1999. Interview by author. February. Stockholm.

Lemmel, Magnus. (Deputy Director General, European Commission, Industrial Affairs [DG III]). 1997. Interview by author. February. Brussels.

Lennon, Peter John. (Advisor, European Commission, Industrial Affairs [DG III]). 1997. Interview by author. February. Brussels.*

Linton, Fredrik. (European Commission, Industrial Affairs [DG III]). 1997. Interview by author. February. Brussels.

———. 1997. Interview by author. June. Brussels.

———. 1997. Interview by author. September. Brussels.

———. 1998. Interview by author. May. Brussels.

List, Ola. (National Defence Research Establishment). 1999. Interview by author. August. Stockholm.

Magnusson, Claes. (Swedish Research EU Liaison Office). 1996. Interview by author. December. Brussels.

———. 1997. Interview by author. February. Brussels.

———. 1997. Interview by author. June. Brussels.

Mannertorn, Annelie. (Industry policy Counselor, Permanent Swedish Representation). 1996. Interview by author. December. Brussels.

Mollard La Bruyere, Yves. (Counselor, European Commission, Service Central de Planification pour les Relations Extérieures). 1998. Interview by author. May. Brussels.

Nette, Olivier. (Counselor, European Commission, External Relations [DG I]). 1997. Interview by author. February. Brussels. 1997

Newman, Nicholas. (Head of Unit, European Commission, Science, Research & Development [DG XII]). 1997. Interview by author. February. Brussels.

Ohlsson, Hans Krister. (Swedish Ministry of Industry and Trade Ministry for industrial affairs). 1999. Interview by author. May Stockholm.

Riegert, Anne. (Expert, European Commission, Industrial Affairs [DG III]). 1998. Interview by author. May. Brussels.

——. 1999. Interview by author. September. Brussels.

Roger, Denis. (Expert, European Commission, Science, Research & Development [DG XII]). 1997. February. Brussels.

——. 1997. Interview by author. June. Brussels

——. 1997. Interview by author. September. Brussels.

Schlieper, Andries. (Chairman, WEAG NADs). 1999. Interview by author. September. Brussels.

Schmitt von Sydow, Helmut. 1997. (Director, European Commission, Industrial Affairs [DG III]). Interview by author. June. Brussels.

Sivertsson, Anders. (Military Attaché, Swedish Embassy). 1997. Interview by author. January. Brussels.

Spagnolli, Alberto. (European Commission, Competition [DG IV]). 1999. Interview by author. September. Brussels.

Sundblad, Helena. (European Commission, Industrial Affairs [DG III]). 1997. Interview by author. February. Brussels.

Svensson, Åke. (General Manager, SAAB AB). 1998. Interview by author. February. Linköping.

von Sydow, Björn. (Swedish defence minister). Interview by author. October. Stockholm.

Theorin, Maj-Britt. (Swedish MEP [Committee on Foreign Affairs, Security and Defence Policy]). 1997. Interview by author. January. Brussels.*

Tjäder, Thomas. (Director Public Affairs, Celsius Corp.). 1998. Interview by author. February. Stockholm.

Törnblom, Dag. (Head of the Association of Swedish Defence Industries). 1997. Interview by author. June. Stockholm.

——. 1998. Interview by author. February. Stockholm.

——. 1998. Interview by author. May. Stockholm.

Törnqvist, Stefan. (Researcher, National Defence Research Establishment, Stockholm [Secretary in the Working Group on Research and Technology within the LOI framework]). 1998. Interview by author. December. Stockholm.

——. 1999. Interview by author. May. Stockholm.

——. 2000. Interview by author. September. Stockholm.

Vasquez, Fernando. (Deputy Head of Unit, European Commission, General Directorate V—Employment, Industrial Relations & Social Affairs [DG V]). 1999. Interview by author. September. Brussels.

Voje, Kristen. (Head of the Information Unit, EUREKA secretariat). 1997. Interview by author. February. Brussels.

Woodcock, Graham. (Secretary-General, European Defence Industries Group [EDIG]). 1999. Interview by author. September. Brussels.

* Telephone Interview

References

Official Documents

AECMA (European Association of Aerospace Industries). 1996. "Towards a European Aerospace Policy—Perspectives and Strategies for the Aerospace Industry." Brussels: AECMA.
——. 2000. Brussels: AECMA.
"Amsterdam Treaty." Treaty of Amsterdam Amending the Treaty on European Union, the Treaties Establishing the European Communities, and Certain Related Acts, as Signed in Amsterdam on 2 October 1997.
Bangemann, Martin. 1996. "The Future of Europe's Defence Industry." Speech at "The Future of Europe's Defence Industry Conference." Brussels, 18 June 1996.
"Brussels Treaty." Treaty of Economic, Social and Cultural Collaboration and Collective Self-Defence, Signed at Brussels on 17 March 1948.
CEPS (Centre for European Policy Studies). 1997. "Defence Equipment Cooperation." *Report of a CEPS Working Party Number 15.*
——. 1999. "Future Cooperation among European Defence Industries in the light of European Multinational Forces."*Report of a CEPS Working Party.* www.cordis.lu/fifth/src/ms-se 1.htm#vie. Accessed January 29 1999.
Declaration of the European Council on Strengthening the Common European Policy on Security and Defence (the Cologne Declaration) 1999. http://ue.eu.int/en/Info/eurocouncil/index.htm. Accessed 13 October 2000.
"Draft Action Plan for the Defence-related Industry." 1997. Directorate-General III, Brussels, 20 August.
EDIG (European Defence Industries Group). 1995. Position paper.

——. 1996. Position Paper.

——. 1997. Position Paper.

——. 1998, Position Paper.

——. 1999. Position Paper.

ESC (Economic and Social Committee of the European Communities) 1997. "Opinion on the Communication from the Commission—The challenges facing the European defence-related industry, a contribution for action at European level," 19-20 March.

European Council, Presidency Conclusions. 1999. Cologne, 4 June. "The Cologne Declaration." Accessed 13 October 2000.

——. 2000. Santa Maria Da Feira 19 and 20 June. http://ue.eu.int/en/Info/ eurocouncil/index.htm. Accessed 13 October 2000.

European Commission. 1990. *Industrial Policy in an Open and Competitive Environment: Guide-lines for a Community Approach* (COM (90) 556 final).

——. 1990. *Commission Opinion of 21 October 1990 on the Proposal for Amendment of the Treaty Establishing the European Economic Community with a View to Political Union* (COM (90) 600 final).

——. 1992. *The European aircraft Industry—First Assessment and Possible Community Actions.* (COM (92) 164 final).

——. 1992. *Proposal for a Council Regulation (EEC) on the Control of Exports of Certain Dual-Use Goods and Technologies and of Certain Nuclear Products and Technologies.* (COM (92) 317 final).

——. 1993. *Growth, Competitiveness, Employment–The Challenges and the Ways Forward into the Twenty-first Century.* (COM (93) 700 final).

——. 1994. *Research and Technological Development – Achieving coordination Through Cooperation.* (COM (94) 438 final).

——.1995. *Green Paper on Innovation.*

——. 1996. *The Challenges Facing the European Defence-Related Industry, A Contribution for Action at European Level.* (COM (96) 10 final).

——. 1996. *The European Union and Space.* (COM (96) 617 final).

——. 1997. *Proposal for a European Parliament and Council Decision Concerning the 5th Framework Programme of the European Community for Research, Technological Development and Demonstration Activities (1998-2002) Proposal for a Council Decision concerning the 5th Framework Programme of the European Atomic Energy Community (Euratom) for Research and Training Activities (1998-2002).* (COM (97) 142 final).

——. 1997. *The European Aerospace Industry Is In Urgent Need of Restructuring– Meeting the Global Challenge.* (COM) (97) 466 final).

——. 1997. *Implementing European Union Strategy on Defence-Related Industries.* (COM (97) 583 final).

——. 1998. *Proposal for a Council Regulation (EC) Setting Up a Community Regime for the Control of Exports of Dual-Use Goods and Technologies.* (COM (98) 257 final).

——. 1998. *Report to the European Parliament and the Council on the Application of Regulation (EC) 3381/94 Setting Up a Community System of Export Controls Regarding Dual-Use Goods.* (COM (98) 258 final).

——. 1999. *First Report by the High Level Group on Benchmarking.* Directorate General III, Industry, No 2.

——. 2002. *A Project for the European Union.* (COM (02) 247 final).

European Court of Justice: 1995a C-70/94: "Judgement of the Court 17 October 1995, Fritz Werner Industrie—Ansrustungen GmbH v Federal Republic of Germany."

——. 1995b. C-83/94: "Judgemement of the Court of 17 October 1995, Criminal proceedings against Peter Leifer, Reinhold Otto Krauskopf and Otto Holzer."

——. 1997. C-124/95: "Judgement of the Court of 14 January 1997, The Queen, ex parte Centr—Com Srl v HM Treasury and Bank of England."

——. 1999. C-414/97: "Judgement of the Court 16 September 1999, Commission of the European Communities v Kingdom of Spain."

European Parliament. 1996. The Committee on Research, Technological Development and Energy, 22 October.

——. 1997. *Defence-related industry Resolution on the Commission Communication on the Challenges Facing the European Defence, a Contribution for Action at European Level.* (A4-0076/97) (Committee on Foreign Affairs, Security and Defence Policy, Rapporteur: Mr Gary Titley).

——. 1998. *Report on the Communication From the Commission on Implementing European Union Strategy on Defence-Related Industries.* (COM (97)0583-C4-0223/98 (A4-0482/98) (Committee on Foreign Affairs, Security and Defence Policy, Rapporteur: Mr Gary Titley).

Ferari, Giovanni Battista. 1995. "NATO's New Standardization Organization Tackles an Erstwhile Elusive Goal." *Nato Review,* 43, 3: 33-35.

Framework Agreement between the French Republic, the Federal Republic of Germany, the Italian Republic, the Kingdom of Spain, the Kingdom of Sweden, and the United Kingdom of Great Britain and Northern Ireland Concerning Measures to Facilitate the Restructuring and Operation of the European Defence Industry. 2000 Farnborough July.

Franco-German Defence and Security Council, Mainz, 9 June 2000.

Helsinki European Council, 10 and 11 December 1999. http://ue.eu.int/en/Info/eurocouncil/index.htm. Accessed 13 October 2000.

Hoon, Geoffrey. 2000. Speech, European Defence Conference. 28 March.

IRDAC (Industrial R&D Advisory Committee of the Commission of the European Communities). 1993. "IRDAC Opinion on Framework Programme IV." 8 October.

"Joint Declaration Issued at the British-French Summit, Saint Malo, France, 3-4 December 1998."

"Joint Statement, April 20, 1998, by the defence ministers of France, Germany, Great Britain, Italy, Spain, and Sweden."

Jospin, Lionel. 1998. Discours à la L'institut des Hautes Etudes de Défense Nationale à Paris 3 September 1998. (Translation by the author).

"Letter of Intent (LOI) between Ministers of Defence from France, Germany, Italy, Spain and Sweden Concerning measures to facilitate the restructuring of European defence industry." July 1998.

Masterplan for the European Armaments Agency. 17 November 1998. WEAG. Brussels.

Ministère de la Défense. 1999. "The French Armaments Policy—Industrial Stakes, European Strategy, Operational Capability." Délégation générale pour l'armement, Paris, France.

Ministry of Defence, UK. 1999. *White Paper*. Available at http://www.mod.uk/publications/whitepaper1999/index.htm. Accessed 13 June 2003.

NAC. 1997. "Conference of National Armaments Directors—NATO Armaments Review with a focus on the CAPS." 17 December (NATO Unclassified).

NAC. 1998a. "Conference of National Armaments Directors (CNAD)—NATO Armaments Review," 16 April 16 1998 (NATO Unclassified).

NAC. 1998b. "Conference of National Armaments Directors—NATO Armaments Review," October 21 (NATO Unclassified).

Naumann, Klaus. 1996. "From cooperation to interoperability." *Nato Review*. 44, 4: 17-20.

OCCAR (Organisme Conjoint de Coopération en Matière d'Armement). 1996. "Arrangement Administratif." Strasbourg le 12 novembre 1996, Strasbourg.

———. 1998. "Convention", Farnborough le 9 septembre 1998.

Official Journal. 1994. No 367, 31/12.

POLARM 3/1998a.

———. 5/1998b.

———. 10/1998c.

———. 11/1998d.

———. 15/1998e.

———. 16/1998f.

———. 2720/1998g.

———. 8421/1998h.

———. 9232/1998i.

———. POLARM 4 June 1998

———. 1999a. "Draft Common Position on Framing a European Armaments Policy.

———. 1999b. 6 May.

Report of the Working Group on Armaments'Exports. LOI. 1999.

———. 1999b. Report of the Working Group on Harmonization of Military Requirements. LOI.

———. 1999c. Report of the Working Group on Research and Technology. LOI

———. 1999d. Report of the Working Group Security of Information. LOI

———. 1999e. Report of the Working Group on Security of Supply. LOI.

———. 1999f. Report of the Working Group on the Treatment of Technical Information. LOI:

Report to Ministers from the Executive Committee under the Letter of Intent of 6 July 1998 concerning Measures to Facilitate the Restructuring of European Defence Industry.

Scharping, Rudolf. 2000a. Speech at NATO Defense College, Rome, January 11.

———. 2000b. Speech at the London School of Economics, March 13.

SCITEC (Study Team established by Defence Ministers of Nations of the Western European Armaments Group with the participation of the European Defence Industry Group) (1998): "Building on Success—WEAG Science and Technology Strategy." The Western European Union. Brussels.

Swedish Government Bill 1998/1999:74: "Förändrad Omvärld—Omdanat Försvar (Changed Global Environment—Changed Defense).

Treaty on European Union.

WEAG. 1999. Executive Summary of the Activities of the GNE on the Masterplan for an EAA, November.

WEU Secretariat General www.weu.int/weag/int/eng/info/weag.htm 16 November 1997.

Other References

Adams, Gordon. 1999. "The Atlantic Option." *Financial Times* 28 January.

Adler, Emmanuel. 1997. "Seizing the Middle Ground." *European Journal of International Relations* 3, no 3: 319-64.

Aggarwal, Vinod K., ed. 1998. *Institutional Designs for a Complex World*. Ithaca/London: Cornell University Press.

Ahrne, Göran. 1994. *Social Organizations*. London: SAGE

Ahrne, Göran and Peter Hedström. 1999. *Organisationer och samhälle—Analytiska perspektiv*. Lund: Studentlitteratur.

Ahrne, Göran and Apostolis Papakostas. 1994. "In the Thick Organizations." In *Social Organizations*, ed. Göran Ahrne. London: SAGE.

Aspinwall, Mark, and Gerald Schneider. 2000. "Same Menu, Separate Tables: The Institutionalist Turn in Political Science and the Study of European Integraton." *European Journal of Political Research* 38: 1-36.

Aviation Week & Space Technology. 1999. 25 October.

———. 2000a. 5 June.

———. 2000b. 24 July.

Axelson, Mattias, and Andrew James. 2000. *The Defence Industry and Globalisation—Challenging Traditional Structures*. Stockholm: The Swedish Defence Research Establishment.

Bainbridge, Timothy (with Anthony Teasdale). 1997. *The Penguin Companion to European Union*. London/New York: Penguin Books.

Barnes, Ian, and Pamela Barnes. 1995. *The Enlarged European Union*. London: Longman Group.

Barnett, Michael N., and Martha Finnemore. 1999. "The Politics, Power and Pathologies of International Organizations." *International Organization* 53, no. 4: 699-732.

Bauer, H. 1992. "Institutional Frameworks for Integration of Arms Production in Western Europe." In *Restructuring of Arms Production in Western Europe*, ed. Michael Brzoska and Peter Lock. Oxford: Oxford University Press.

Berger, Peter, and Thomas Luckmann. 1967. *The Social Construction of Reality*. New York: Doubleday.

Bergström, Göran, and Kristina Boréus. 2000. *Textens mening och makt— Metodbok i samhällsvetenskaplig textanalys*. Lund: Studentlitteratur.

Bernitz, Ulf. 2001. "Flexibel integration—väg för EU:s utveckling?" In *Europaperspektiv*, ed. Ulf Bernitz, Sverker Gustavsson, and Lars Oxelhiem. Stockholm: Santérus.

Blitz, James, and Alexander Nicoll. 2000. "BAe and EADS Scramble to Make an Italian Ally." *Financial Times*, 4 February.

Boli, John, and George M. Thomas, ed. 1999. *Constructing World Culture*. Stanford: Stanford University Press.

Börzel, Tania and Tomas Risse. 2000. "Who Is Afraid of European Federation? How to Constitutionalise a Multi-Level Governance System." *Jean Monnet Working Paper No 7, Symposium: Responses to Joschka Fischer*, Florence: European University Institute.

Bourdieu, Pierre. 1996. *The Rules of Art—Genesis and Structure of the Literature Field*. Cambridge: Polity Press.

Bourdieu, Pierre, and Loïc J. D. Wacquant. 1992. *An Invitation to Reflexive Sociology*. Chicago: University of Chicago Press.

Bretherton, Charlotte, and John Vogler. 1999. *The European Union as a Global Actor*. London: Routledge.

Britz, Malena. 2000. "The Development of Swedish and French Defence Industrial Companies 1994-1999—A Comparative Study." *SCORE Working Paper* 6.

Britz, Malena, and Arita Eriksson. 2000. *British, German and French Defence Industrial Policy in the 1990s*. Stockholm: Swedish Research Defence Establishment.

Broady, Donald, ed. 1998. *Kulturens Fält—en antologi (The Fields of Culture— An Anthology)*. Göteborg: Daidalos.

Brunsson, Nils. 1998. "Homogeneity and Heterogeneity in Organizational Forms as the Result of Cropping-Up Processes." In *Organizing Organizations*, ed. Nils Brunsson and Johan P. Olsen. Bergen: Fagbokforlaget.

———. 1999. "Standardization as Organization." In *Organizing Political Institutions*, ed. Morten Egeberg and Per Laegreid. Oslo: Scandinavian University Press.

Brunsson, Nils, and Bengt Jacobsson, ed. 2000. *A World of Standards*. Oxford: Oxford University Press.

Brunsson, Nils and Johan P. Olsen. 1998. "Organization Theory: Thirty Years of Dismantling, and Then?" In *Organizing Organizations*, ed. Nils Brunsson and Johan P. Olsen. Bergen: Fagbokforlaget.

Brunsson, Nils, and Kerstin Sahlin-Andersson. 1997. "Constructing Organizations." *SCORE Working Paper 8*.

Brzoska, Michael. 2000. "The Future of Defense Production—Europe's Chances, Choices and Conduct." Draft paper presented at Workshop: Defense Industry Restructuring in Europe in the 1990s and Beyond, SIPRI, Stockholm, 13 September.

Bulmer, Simon. 1994. "The New Governance of the European Union: A Neo-Institutionalist Approach." *Journal of European Public Policy* 13, no. 4: 351-80.

———. 1998. "New Institutionalism and the Governance of the Single European Market." *Journal of European Public Policy* 5, no. 3: 365-86.

Cable, Vincent. 1995. "What is International Security?" *International Affairs* 2: 305-24.

Caporaso, James A. 1996. "The European Union and Forms of State. Westphalian, Regulatory or Post-Modern?" *Journal of Common Market Studies* 34: 29-51.

Carvalho, Abraao. 2000. "A European Perspective of Defence Industry Matters." In *The Changing European Defence Industry Sector—Consequences for Sweden?* ed. Arita Eriksson and Jan Hallenberg. Stockholm: Swedish National Defence College.

Cederman, Lars-Erik. 2001. "Political Boundaries and Identity Trade-Offs." In *Constructing Europe's Identity—The External Dimension*, ed. Lars-Erik Cederman. London: Lynne Rienner.

CEPS (Centre for European Policy Studies). 1996. "Defence Equipment Cooperation."*Report of a CEPS Working Party Number 15*.

———. 1999: "Future Cooperation among European Defence Industries in the Light of European Multinational Forces." *Report of a CEPS Working Party*.

Chapman, Peter. 2000. "Gibraltar Accord Set To Unblock Raft of Frozen Union Proposals." *European Voice* 13 April.

Christiansen, Thomas. 1997. "Reconstructing European Space: From Territorial Politics to Multilevel Governance." In *Reflective Approaches to European Governance*, ed. Knud-Erik Jørgensen. London: Macmillan.

Christensen, Tom, and Per Laegreid. 1998. "Public Administration In a Democratic Context—A Review of Norwegian Research." In *Organizing Organizations*, ed. Nils Brunsson and Johan P. Olsen. Bergen: Fagbokforlaget.

Cini, Michelle, and Lee McGowan. 1998. *Competition Policy in the European Union*. London: Macmillan.

Cordes, Renée. 2000. "Giant Defence Merger Set for Clearance." *European Voice*. 4 May.

Cornish, Paul. 1997. *Partnership in Crisis: the US, Europe and the Fall and Rise of NATO*. London: Royal Institute of International Affairs.

Coss, Simon. 2000. "Spanish Poll to Delay Talks on EU Firms." *European Voice*. 3 February.

Cram, Laura. 1997. *Policy-Making in the EU*. London: Routledge.

Crawford, Beverly. 1994. "The New Security Dilemma under International Economic Interdependence." *Millennium* 23, no. 1: 25-56.

——. 1995. "Hawks, Doves, but No Owls: International Economic Interdependence and Construction of the New Security Dilemma." In *On Security*, ed. Ronnie Lipschutz. New York: Columbia University Press.

Cresson, Edith. 1996. "Interview." *Innovation & Technology Transfer.* February.

Davies, Ian. 2000. "Regulation of European Arms and Dual-Use Exports in a Transnational Defence Industrial Environment: The EU Code of Conduct on Arms Exports." Paper presented at SIPRI at COST A10 workshop, 13-15 October, Stockholm.

de Briganti, Giovanni.1996. "Europe Weighs Blanket Arms Agency." *Defense News*. 14 October.

——. 1997a. "Can Italy, Franc Build Ties To Bind?" *Defense News*. 29 September.

——. 1997b. "Weu Ends Yearlong Limbo Of Arms Agency," *Defense News*. 17 November.

Defense Daily.

Delhotte, Pierre. 1999. Letter to the Author, November 30.

——. 2000. "The Western European Armaments Group (WEAG)." In *The Changing European Defense Industry Secctor—Consequences for Sweden*, ed. Arita Eriksson and Jan Hallenberg. Stockholm: Swedish National Defense College.

Den Boer, Monica and William Wallace. 2000. "Justice and Home Affairs —Integration Through Incrementalism." In *Policy—Making in the European Union*, fourth ed., ed. Helen Wallace and William Wallace. Oxford: Oxford University Press.

De Spiegeleire, Stephan. 1999. "The European Security and Defence Identity and NATO: Berlin and Beyond." In *European Security Integration*, ed. Mattias Jopp and Hanna Ojanen. Helsinki:Ulkopoliittinen instituutti/Bonn: Institut fur Europäische Politik/Paris: WEU Institute for Security Studies.

DiMaggio, Paul J. 1983. "State Expansion and Organizational Fields." In *Organizational Theory and Public Policy*, ed. Richard H. Hall and Robert E. Quinn. London: SAGE.

DiMaggio, Paul J. and W.W. Powell. 1991. "The Iron Cage Revisited: Institutional Isomorphism and Collective Rationality In Organizational Fields." In *The New Institutionalism in Organizational Analysis*, ed. Walter W. Powell and Paul J. DiMaggio. Chicago/London: University of Chicago Press.

Ehrhart, Hans-Georg. 2002. "What Model for CFSP?" *Chaillot Papers No 55,*

Paris: Institute for Security Studies, European Union.

Eising, Rainer and Beate Kohler-Koch, ed. 1999. *The Transformation of Governance in the European Union.* London: Routledge.

Ekengren, Magnus. 1998. *Time and European Governance.* Ph.D. diss. Stockholm: Department of Political Science, Stockholm University.

Eliassen, Kjell, ed. 1998. *Foreign and Security Policy in the European Union.* London: SAGE.

Eriksson, Arita and Jan Hallenberg, eds. 2000. *The Changing European Defence Industry Sector—Consequences for Sweden?* Stockholm: Swedish National Defence College.

European Voice, 1999. "15 November General Affairs Council/Defence Ministers. 18 November.

Fierke, K. and Antje Wiener. 1999. "Constructing Institutional Interests: EU and NATO Enlargment." *Journal of European Public Policy* 6: 721-42.

Financial Times. 1999a. "Dasa Chief's Hopes Fade for Pan-European Defence." January 25.

——. 1999b. "Pledge on European Military Capability." 31 May.

——. 1999c. "Leader: Future Isn't Plain." 15 October.

——. 1999d. "EADS Seeks Italian Alliance." 30 December.

——. 2000a. "Eurofighter Missile Companies in Plea to Pentagon over Supplying Meteor Weapons." 18 February.

——. 2000b. "DASA Prepares for Ned Over Aerospace Merger." 9 March.

——. 2000c. "Italy Chooses EADS for Aircraft Link-Up." 14 April.

——. 2000d. "UK Defence Orders Go To Europe." 17 May.

——. 2000e. "Chirac Calls for European Rapid Reaction Force." 31 May.

Fligstein, Neil, and Jason McNichol. 1998. "The Institutional Terrain of the European Union." In *European Integration and Supranational Governance*, ed. Wayne Sandholtz and Alec Stone Sweet. London: Oxford University Press.

Forssell, Anders. 1992. *Moderna Tider i Sparbanken*, Stockholm: Nerenius & Santérus.

Forster, Anthony, and William Wallace. 2000. "Common Foreign and Security Policy—From Shadow to Substance?" In *Policy-Making in the European Union*, ed. Helen Wallace and William Wallace. Oxford: Oxford University Press.

Foucault, Michel. 1972. *The Archaeology of Knowledge.* New York: Pantheon Books.

Gamson, William. 1988. "Political Discourse and Collective Action." In *From Structure to Action: Comparing Social Movement Research Across Cultures*, ed. Bert Klandermans, Hanspeter Kriesi, and Sidney Tarrow. Greenwich, Conn./London: JAI Press.

Garten, Jeffrey. 1992. *A Cold Peace: America, Japan, Germany and the Struggle for Supremacy.* New York: Times Books.

George, Alexander, and Timothy McKeown. 1985. "Case Studies and Theories of Organizational Decision Making." In *Advances in Information Processing in Organizations*, vol 2, ed. Lee Sproull and V. Larkey. Greenwich, Conn.: JAI Press.

Gibson, D., and Robert Goodin, R.(1999). "The Veil of Vagueness: A Model of Institutional Design." In *Organizing Political Institutions*, ed. Morten Egeberg and Per Laegreid. Oslo: Scandinavian University Press.

Giddens, Anthony. 1979. *Central Problems in Social Theory: Action, Structure and Contradiction in Social Analysis*. Berkeley: University of California Press.

———. 1984. *The Constitution of Society. Outline of the Theory of Structuration.* Cambridge: Polity Press.

Gissler, Lars. 2000. Speech. Conference at SIPRI. 13 October

Goffman, Erving. 1974. *Frame Analysis*. Boston: Northeastern University Press.

Green Cowles, Maria, James A. Caporaso and Tomas Risse, eds. 2001. *Transforming Europe: Europeanization and Domestic Change*. Ithaca, N.Y.: Cornell University Press.

Grosser, Alfred. 1980. *The Western Alliance—European-American Relations since 1945*. London: Macmillan.

Gummett, Philip, and Josephine Anne Stein. 1997. *European Defence Technology in Transition*. Amsterdam: Harwood Academic Publishers.

Guzzetti, Luca. 1995. *A Brief History of European Union Research Policy*. Brussels: European Commission.

Haaland Matláry, Janne. 2002. *Intervention for Human Rights in Europe*. Houndmills: Palgrave.

Haas, Ernst. 1958/68. *The Uniting of Europe: Political, Social, and Economic Forces 1957-57*. Stanford, Calif.: Stanford University Press.

———. 1964. *Beyond the Nation-State*. Stanford, Calif.: Stanford University Press.

Haas, Peter. 1992. "Introduction: Epistemic Communities and International Policy Coordination." *International Organization* 46: 1-35.

Hagelin, Björn. 1997. "Sweden." In *European Defence Technology in Transition*, ed. Philip Gummett and Josephine Anne Stein. Amsterdam: Harwood Academic Publishers.

———. 1999. "Sweden's Search for Military Technology." In *The Restructuring of Arms Production in Western Europe,* ed. Michael Brzoska and Peter Lock. Oxford: Oxford University Press/SIPRI.

———. 2000. "Swedish for how long? The Nation's Defence Industry in an International Context." In *The Changing European Defence Industry Sector—Consequences for Sweden?* ed. Arita Eriksson and Jan Hallenberg. Stockholm: Swedish National Defence College.

Hall, Peter. 1993. "Policy Paradigms, Social Learning and the State." *Comparative Politics* 25: 275-96.

Hart, Jeffrey. 1992. *Rival Capitalists—International Competitiviness in the United*

States, Japan and Western Europe. Ithaca, N.Y.: Cornell University Press.

Hasenclever, Andreas, Peter Mayer, and Volker Rittberger. 1997. *Theories of International Regimes.* Cambridge: Cambridge University Press.

Hayward, Keith. 1997. "Towards a European Weapons Procurement Process." *Chaillot Papers 27.* Paris: Institute for Security Studies, Western European Union.

Heisbourgh, François et al. 2000 "European Defence: Making It Work." *Chaillot Papers 40.* Paris: Institute for Security Studies, Western European Union.

Héritier, Adrienne. 1997. "Policy-making by Subterfuge: Interest Accommodation, Innovation and Substitute Policy Areas." *Journal of European Public Policy* 4, no. 2: 171-89.

——. 1999. *Policy-Making and Diversity in Europe—Escape from Deadlock.* Cambridge: Cambridge University Press.

——, ed. 2001. *Common Goods: Reinventing European Integration Governance.* Lanham, Md.: Rowman & Littlefield.

Hitchens, Theesa, and Brooks Tigner. 1998. "Clashing Agendas Threaten Europe Arms Unity." *Defense Daily.* 30 March.

Hix, Simon. 1994. "The Study of the European Community: The Challenge to Comparative Politics." *West European Politics* 17, no. 1: 1-30.

——. 1996. "CP, IR and the EU! A Rejoinder to Hurrell and Menon." *West European Politics* 19, no. 4: 802-4.

Hodges, Michael. 1983. "Industrial Policy: Hard Times or Great Expectations?" In *Policy Making in the European Community*, ed. Helen Wallace, William Wallace and Carol Webb. London: John Wiley & Sons.

Hooghe, Lisbeth, and Gary Marks. 2001. *Multi-Level Governance and European Integration.* Lanham, Md: Rowman & Littlefield.

Jachtenfuchs, Markus. 1996. *International Policy-Making as a Learning Process?* Aldershot: Ashgate.

Jacquemin, Alexis, and Lucio R. Pench, eds. 1997. *Europe Competing in the Global Economy—Reports of the Competitiveness Advisory Group.* Cheltenham: Edward Elgar.

Jakubyszyn, Christophe and Anne-Marie Rocco. 2000. "EADS Et Ses 96 000 Salariés en Quête D'Un Modèle Européen D'Entreprise." *Le Monde*, 25 February.

James, Andrew. 1999. *Post-Merger Strategies of the Leading US Defence Aerospace Companies.* Research Report. Stockholm: Swedish Defence Research Establishment.

Jerneck, Magnus. 2001. "Är flexibel integration i EU en bärkraftig princip?" In *Europaperspektiv*, ed. Ulf Bernitz, Sverker Gustavsson and Lars Oxelheim. Stockholm: Santérus.

Jopp, Mattias and Hanna Ojanen, eds. 1999. *European Security Integration.* Helsinki: Ulkopoliittinen instituutti/Bonn: Institut fur Europäische Politik/

Paris: WEU Institute for Security Studies.

Josefsson, Lars. 2000. Statement. 5th Forum Europa Defense Industries Conference. 23 May. Brussels.

Keohane, Robert and Joseph S. Nye, Jr. 1977. *Power and Interdependence.* Boston/Toronto: Little Brown.

Kingdon, John. 1984. *Agendas, Alternatives, and Public Policies.* Boston: Little, Brown.

Kohler-Koch, Beate. 1997. "Organized Interests in European integration: The Evolution of a New Type of Governance?" In *Participation and Policy-Making in the European Union,* ed. Helen Wallace and Alasdair Young. Oxford: Clarendon Press.

——. 1999. "The Evolution and Transformation of European Governance." In *The Transformation of Governance in the European Union,* ed. Beate Kohler-Koch and Rainer Eising. London: Routledge.

Kohler-Koch, Beate and Rainer Eising. 1999. "Introduction: Network Governance in the European Union." In *The Transformation of Governance in the European Union,* ed. Beate Kohler-Koch and Rainer Eising. London: Routledge.

Krasner, Stephen, ed. 1983. *International Regimes.* Ithaca, N.Y.: Cornell University Press.

——. 1999. *Sovereignty— Organized Hypocrisy.* Princeton: Princeton University Press.

Kratochwil, Friedrich V. 1989. *Rules, Norms and Decisions.* New York: Cambridge University Press.

Laegreid, Per, and Paul Roness. 1999. "Administrative Reform as Organized Attention." In *Organizing Political Institutions,* ed. Morten Egeberg and Per Laegreid. Oslo: Scandinavian University Press.

Larsen, Peter Thal. 1999. "Kosovo Conflict Highlights Real Winners in Wars." *Financial Times.* June 1.

Lauman, Edward O., and David Knoke. 1987. *The Organizational State: Social Choice in National Policy Domains.* Madison: University of Wisconsin Press.

Lequesne, Christian. 2000. "The European Commission: A Balancing Act between Autonomy and Dependence." In *European Integration after Amsterdam: Institutional Dynamics and Prospects for Democracy,* ed. Karlheinz. Neunreither and Antje Wiener. Oxford: Oxford University Press.

Lindberg, Leon N. 1963. *The Political Dynamics of European Economic Integration.* Stanford, Calif.: Stanford University Press.

Lindberg, Leon N. and Stuart A. Scheingold. 1970. *Europe's Would-Be Polity: Patterns of Change in the European Community.* Englewood Cliffs, N.J.: Prentice Hall.

Lindgren, Fredrik. 1998. *Interoperability As a Factor in Armaments Collaboration.* Stockholm: Swedish Defence Research Establishment.

Lipschutz, Ronnie, ed. 1995. *On Security.* New York: Columbia University Press.

Lovering, John. 2000. "The US Influence on European Defence Industry Restructuring." Paper presented at COST A10 Workshop at SIPRI, Stockholm 13-15 October..

Magaziner, Ira and Mandy Patinkin. 1990. *The Silent War: Inside the Global Business Battles Shaping America's Future.* New York: Viking Books.

Majone, Giandomenico. 1997. "The New European Agencies: Regulation by Information." *Journal of European Public Policy* 4, no. 2: 262-75.

March, James A. 1981. "Decision Making Perspective." In *Perspectives on Organization Design and Behavior*, ed. Andrew H. Van de Ven and William F. Joyce. New York: Free Press.

———. 1994. *A Primer on Decision Making: How Decisions Happen.* New York: Free Press.

March, James A. and Johan P. Olsen. 1989. *Rediscovering Institutions.* New York: Free Press.

———. 1998. "The Institutional Dynamics of International Political Orders." *International Organization* 52, no. 4: 943-70.

Marcussen, Martin, Thomas Risse, Daniela Engmann-Martin, Hans Joachim Knopf, and Klaus Roscher. "Constructing Europe? The Evolution of French, British, and German Nation State Identies." *Journal of European Public Policy* 6, no. 4: 614-633.

Marks, Gary. 1993. "Structural Policy and Multilevel Governance in the EC. " In *The State of the European Community II: The Maastricht Debates and Beyond*, ed. Alan Cafruny and Glenda Rosenthal. Boulder, Colo.: Lynne Rienner.

Marks, Gary, Lisbeth Hooghe, and Kermit Blank. 1996. "European Integration from the 1980s: State Centric v. Multi-Level Governance." *Journal of Common Market Studies* 34, no. 3: 341-78.

Mastanduno, Michael. 1992. *Economic Containment: CoCom and the Politics of East-West Trade.* Ithaca, N.Y.: Cornell University Press, 1992.

Merton, Robert K. 1965. *On the Shoulders of Giants.* New York: The Free Press.

Meyer, John, and Brian Rowan. 1977/91. "Institutionalized Organizations: Formal Structures as Myth and Ceremony." In *The New Institutionalism in Organizational Analysis*, ed. Walter Powell and Paul DiMaggio. Chicago: University of Chicago Press.

Mezzadri, Sandra. 2000. "L'overture des marchés de la défense: enjeux et modalités." Paris: Western European Union Institute for Security Studies, *Publications Occasionnelles* 12.

Middlemaas, Keith. 1995. *Orchestring Europe—The Informal Politics of the European Union 1973-95.* London: Fontana Press.

Missiroli, Antonio. 1999. "Towards a European Security and Defense Identity? Record—State of Play—Prospects." In *European Security Integration*, ed. Mattias Jopp and Hanna Ojanen. Helsinki: Ulkopoliittinen instituutti/Bonn: Institut fur Europäische Politik/Paris: WEU Institute for Security Studies.

———. 2000. "CFSP, Defence and Flexibility." *Chaillot Papers 38.* Paris: Institute for Security Studies, Western European Union.

Moravcsik, Andrew. 1991. "Negotiating the Single European Act." *International Organization* 45, no. 1: 19-56.

——. 1993. "Preferences and Power in the European Community: A Liberal Intergovernmentalist Approach." *Journal of Common Market Studies* 31, no. 4: 473-524.

——. 1998. *The Choice for Europe: Social Purpose and State Power from Messina to Maastricht*. Ithaca, N.Y.: Cornell University Press.

Morrocco, John O. 2000a. "U.K.'s Tilt Toward Europe Poses Multiple Challenges." *Aviation Week&Space Technology.* 22 May.

——. 2000b. "EADS, Northrop Grumman Broaden Cooperative Links." *Aviation Week&Space Technology.* 12 June.

Mörth, Ulrika. 1996. *Vardagsintegration—La vie quotidienne—i Europa. Sverige i EUREKA och EUREKA i Sverige (Everyday Integration—La vie quotidienne—Sweden in EUREKA and EUREKA in Sweden)* Ph.D. diss. Stockholm: Department of Political Science, Stockholm University.

——. 1998. "EU Policy-Making in Pillar One and a Half—The European Commission and Economic Security," *Research Report, Swedish Agency for Civil Emergency Planning/SCORE.*

——. 2000a. "Competing Frames in the European Commission—The Case of the Defence Industry/Equipment Issue." *Journal of European Public Policy* 7, no. 2: 173-89.

——. 2000b. "Swedish Industrial Policy and Research and Technological Development: The Case of European Defence Equipment." In *Sweden and EU Evaluated,* ed. Lee Miles. London: Cassel.

——. 2003. "Europeanisation as Interpretation, Translation and Editing of Public Policies." In *The Politics of Europeanisation: Theory and Analysis,* ed. Kevin Featherstone and Claudio Radaelli. Oxford: Oxford University Press.

Mörth, Ulrika, and Bengt Sundelius. 1993. "Dealing With a High Technology Vulnerability Trap: The U.S., Sweden and Industry." *Cooperation and Conflict,* 28, no. 3: 303-28.

Muradian, Vago. 1999. "Weston: EADC May be Stalled, But Lessons Learned Live on." *Defense Daily* 10 August.

Nau, Henry. 1975. "Global Responses to R&D problems in Western Europe: 1955-58 and 1968-1973." *International Organization* 29: 616-54.

Nicoll, Alexander. 1991. "America in Its Sights: BAe Systems Needs a US Deal to Fulfil Its Aim of Becoming the Defence Industry Leader." *Financial Times.* 14 December.

North, Douglass C. 1990. *Institutions, Institutional Change and Economic Performance*. Cambridge: Cambridge University Press.

Nye, Joseph S., Jr. 1990. *Bound to Lead—The Changing Nature of American Power*. New York: Basic Books.

Nye, Joseph S., Jr., and William Owens. 1996. "America's Information Edge."

Foreign Affairs 75, no. 2: 20-36.

Öberg, Ulf .1992. *Om EG-rätten, medlemsstaternas försvarsindustri och nationella säkerhetsintressena.* Stockholm: Swedish Research Defence Establishment.

Pehrson, Lennart. 1999. "Kan aktiebolag flytta?" In *Europaperspektiv,* ed. Ulf Bernitz, Sverker Gustavsson, and Lars Oxelhiem. Stockholm: Nerenius & Santérus.

Peters, Guy. 1994. "Agenda-Setting in the European Community." *Journal of European Public Policy* 1, no. 1: 9-26.

———. 1999. *Institutional Theory in Political Science.* London/New York: Pinter.

Peterson, John. 1992. *The Politics of European Technological Collaboration. An Analysis of the Eureka Initiative.* Ph.D. diss. London School of Economics and York University.

———. 1995. "Decision-Making in the European Union: Towards a Framework for Analysis." *Journal of European Public Policy* 2, no.1: 69-93.

Peterson, John, and Elizabeth Bomberg. 1998. *Decision-making in the European Union.* London: Macmillan.

Peterson, John, and Margaret Sharp. 1998. *Technology Policy in the European Union.* London: Macmillan.

Peterson, John, and Helene Sjursen. 1998. *A Common Foreign Policy for Europe?* London: Routledge.

Pierre, Jon, ed. 2000. *Debating Governance.* Oxford: Oxford University Press.

Premfors, Rune. Forthcoming. "Democracy in Sweden: A Historical and Comparative Perspective".

Prévot, Maurice. 1998. "OCCAR—mode d'emploi." *L'Armement,* March 61: 95-99.

Rawls, John. 1955. "Two Concepts of Rules." *Philosophical Review* 44: 3-32.

Rees, G. Wyn. 1998. *The Western European Union at the Crossroads—Between Trans-Atlantic Solidarity and European Integration.* Boulder, Colo.: Westview Press.

Rhodes, Carolyn, ed. 1998. *The European Union in the World Community.* London/Boulder, Colo.: Lynne Rienner.

Rhodes, Roderick Arthur William. 1997. *Understanding Governance.* Buckingham: Open University Press.

Rohde, Joachim, and Jens van Scherpenberg. 1996. "European Commission/DG I Seminars on Economic Security, 1st Seminar: Technology Trends—The Security/Economic Challenge," Stiftung Wissenschaft und Politik (SWP).

Rosamond, Ben. 1999. "Discourses of Globalisation and the Social Construction of European Identities." *Journal of European Public Policy* 6, no. 4: 652-68.

———. 2000. *Theories of European Integration.* London: Macmillan.

Rosenau, James. 1992. "Governance, Order and Change in World Politics." In *Governance Without Government: Order and Change in World Politics,* ed. James Rosenau and Ernst-Otto Czempiel. Cambridge: Cambridge University

Press.

Rosenau, James and Ernst-Otto Czempiel, eds. 1992. *Governance Without Government: Order and Change in World Politics*. Cambridge: Cambridge University Press.

Ruggie, John Gerard. 1975. "International Responses to Technology: Concepts and Trends." *International Organization* 29: 557-83.

——. 1993. "Territoriality and beyond. Problematizing Modernity in International Relations." *International Organization* 47: 139-73.

——. 1998a. "What Makes the World Hang Together? Neo-utilitarianism and the Social Constructivist Challenge." *International Organization* 52, no. 4: 855-85.

——. 1998b. *Constructing the World Polity*. London/New York: Routledge.

Sabatier, Paul, ed. 2000. *Theories of the Policy Process*. Boulder, Colo.: Westview Press.

Sahlin-Andersson, Kerstin. 1986. *Beslutsprocessens komplexitet (The Compexity of the Decisionmaking Process)*. Lund: Studentlitteratur.

——. 1998. "The Social Construction of Projects. A Case Study of Organizing of an Extraordinary Building Project—the Stockholm Globe Arena." In *Organizing Organizations*, ed. Nils Brunsson and Johan P. Olsen. Bergen: Fagbokforlaget.

Sandholtz, Wayne. 1992. *High-Tech Europe—The Politics of International Cooperation*. Berkeley/Los Angeles/Oxford: University of California Press.

Sandholtz, Wayne, Michel Borrus, John Zysman, Ken Conca, Jay Stowsky, Steven Vogel, and Steven Weber. 1992. *The Highest Stakes—The Economic Foundations of the Next Security System*. Oxford: Oxford University Press.

Sandholtz, Wayne, and Alec Stone Sweet. 1998. *European Integration and Supranational Governance*. Oxford: Oxford University Press.

Sandholtz, Wayne, and John Zysman. 1989. "'1992' Recasting the European Bargain." *World Politics* 27, no. 1: 95-128.

Scharpf, Fritz. 1999. *Governing in Europe—Effective and Democratic?* Oxford: Oxford University Press.

Schmitt, Burkard. 2000a. "EADC Is Dead—Long Live EADS!" *Newsletter*, no 28, Institute for Security Studies, Western European Union, Paris.

——. 2000b. "Task Force: 'European Armaments Sector' Fourth Session, 'Towards a Common European Demand for Defence Goods.'" Institute for Security Studies, Western European Union, Paris, January.

——. 2000c. "From Cooperation to Integration: Defence and Aerospace Industries in Europe." *Chaillot Papers 40*, Institute for Security Studies, Western European Union, Paris.

Schön, Donald A., and Martin Rein. 1994. *Frame Reflection*. New York: Basic Books.

Scott, W. Richard. 1995. *Institutions and Organizations*. London: SAGE.

————. 1998. *Organizations: Rational, Natural and Open Systems*. Englewood Cliffs, N.J.: Prentice Hall.

Scott, W. Richard, and John Meyer et al. 1994. *Institutional Environments and Organizations*. Thousand Oaks, Calif.: SAGE.

Searle, John R. 1994. *The Construction of Social Reality*. London: Allen Lane.

Servan-Schreiber, Jean-Jacques. 1968. *The American Challenge*. London: Hamilton.

Sharp, Margaret, and Shearman, Claire. 1987. "European Technological Collaboration." *Chatham House Papers* no 36, London.

Shaw, Jo. 1996. *Law of the European Union*. London: Macmillan.

Shepsle, Kenneth A. 1989. "Studying Institutions: Lessons from the Rational Choice Approach." *Journal of Theoretical Politics* 1: 131-47.

SIPRI 1999. *Yearbook 1999*. Oxford: Oxford University Press for SIPRI.

Sjursen, Helene. 1998. "Missed Opportunity or Eternal Fantasy?: The Idea of European Security and Defence Policy." In *A Common Foreign Policy for Europe?* ed. John Peterson and Helene Sjursen. London: Macmillan.

Sloan, Stanley. 1985. *NATO's Future—Toward a New Transatlantic Bargain*. Washington, D.C.: National Defense University Press.

Smith, Karen Elizabeth. 1998. "The instruments of European Union Foreign Policy." In *Paradoxes of European Foreign Policy*, ed. Jan Zielonka. The Hague: Kluwer Law International.

Smith, Michael. 1996. "The European Union and a Changing Europe: Establishing the Boundaries of Order." *Journal of Common Market Studies* 34: 5-28.

Snow, David A., and Robert D. Benford. 1988. "Ideology, Frame Resonance and Participant Mobilization." In *From Structure to Action. Comparing Social Movements across Cultures*, ed. Bert Klandermans Hanspeter Kriesi and Sidney Tarrow. Greenwich, Conn.: JAI Press Inc.

Soetendorp, Ben. 1999. *Foreign Policy in the European Union*. London: Longman.

Sparaco, Pierre. 2000a. "Streamlining Lends Credence To Finmeccanica/EADS Link." *Aviation Week&Space Technology*. 24 April.

————. 2000b."Accelerating Mergers in Europe Sharpen Consolidation Picture." *Aviation Week&Space Technology*. 25 October.

Sperling, James, and Emil Kirchner. 1997. *Recasting the European Order: Security Architectures and Economic Cooperation*. Manchester: Manchester University Press.

Stern, Charlotta. 1999. "Nyinstitutionell organisationsteori (New-Institutional Organization Theory)" In *Organisationer och samhälle—Analytiska perspektiv(Organizations and Society-Analytical Perspectives)*, ed. Göran Ahrne and Peter Hedström. Lund: Studentlitteratur.

Strange, Susan. 1992. "States, Firms and Diplomacy." *International Affairs* 68:

1-16.

Stråth, Bo, ed. 2000. *Europe and the Other and Europe as the Other*. Bruxelles: Peter Lang.

Stubb, Alexander C.-G. 2000. "Negotiating Flexible Integration in the Amsterdam Treaty." In *European Integration after Amsterdam—Institutional Dynamics and Prospects for Democracy*, ed. Kalrheintz Neunreither and Antje Wiener. Oxford: Oxford University Press.

Sundelius, Bengt, ed. 1989. *The Committed Neutral: Sweden's Foreign Policy*. Boulder, Colo.: Westview Press.

Taft, Willaim H., and William Taylor. 1992. "A Defence Market for NATO." *NATO's Sixteen Nations* 2: 8-11.

Taylor, Paul. 1997. *The European Union in the 1990's*. Oxford: Oxford University Press.

Taylor, Paul, and Peter Schmidt. 1997. "The Role of the Armaments Industry in Supporting the Preparation and Conduct of Military Operations." Ebenhausen: Stiftung Wissenschaft und Politik.

Taylor, Simon. 1999. "EU Leaders Advance on Common Defence Policy." *European Voice* 26 May.

The Economist. 2000. "Sir Janus." December 21.

Thelen, Kathleen and Svein Steinmo. 1992. "Historical Institutionalism in Comparative Politics." In *Structuring Politics. Historical Institutionalism in Comparative Analysis*, ed. Svein Steinmo, Kathleen Thelen and Frank Longstreth. Cambridge: Cambridge University Press.

Thurow, Lester. 1992. *Head to Head: The Coming Economic Battle among Japan, Europe and America*. New York: Morrow.

Tyson, Laura. 1992. *Who's Bashing Whom? Trade Conflict in High Technology Industries*. Washington, D.C.: Institute for International Economics.

Wallace, Helen. 2000. "Flexibility: A tool of Integration or a Restraint on Disintegration?" In *European Integration After Amsterdam—Institutional Dynamics and Prospects for Democracy*, ed. Karlheintz Neunreither and Antje Wiener. Oxford: Oxford University Press.

Wallace, Helen and William Wallace, eds. 2000. *Policy Making in the European Union*. Oxford: Oxford University Press.

Wyatt-Walter, Holly. 1995. "Globalisation, Corporate Identity and European Technology Policy." *Journal of European Public Policy* 2: 427-46.

Yost, David. 1998. *NATO Transformed—The Alliance's New Roles in International Security*. Washington, D.C.: United States Institute of Peace Press.

Zielonka, Jan. 2001. "How New Enlarged Borders will Reshape the European Union." *Journal of Common Market Studies* 3: 507-36.

Zielonka, Jan. 2002. "Introduction." In *Europe Unbound—Enlarging and Reshaping the Boundaries of the European Union*, ed. Jan Zielonka. London: Routledge.

Index

About the Author

Ulrika Mörth is associate professor and lecturer at the Department of Political Science, Stockholm University, and senior researcher and research coordinator at the Stockholm Center for Organizational Research (SCORE).